PLUTO'S REPUBLIC

PETER MEDAWAR

Incorporating

THE ART OF THE SOLUBLE

and

INDUCTION AND INTUITION IN

SCIENTIFIC THOUGHT

Oxford   New York

OXFORD UNIVERSITY PRESS

1984

Oxford University Press, Walton Street, Oxford OX2 6DP
London Glasgow New York Toronto
Delhi Bombay Calcutta Madras Karachi
Kuala Lumpur Singapore Hong Kong Tokyo
Nairobi Dar es Salaam Cape Town
Melbourne Auckland
and associated companies in
Beirut Berlin Ibadan Mexico City Nicosia

Oxford is a trade mark of Oxford University Press

First published 1982
Reprinted 1983
First issued as an Oxford University Press paperback 1984

British Library Cataloguing in Publication Data
Medawar, P. B.
Pluto's republic. – (Oxford paperbacks)
1. Science – Philosophy
I. Title
501   Q175
ISBN 0–19–283039–2

Library of Congress Cataloging in Publication Data
Medawar, P. B. (Peter Brian), 1915–
Pluto's Republic.
(Oxford paperbacks)
'Incorporating The art of the soluble and
Induction and intuition in scientific thought.'
'Oxford University Press paperback' – P.
Bibliography: p.
Includes index.
1. Science – Philosophy – Addresses, essays, lectures.
I. Title. II. Series.
Q175.3.M4 1984 501 83–13234
ISBN 0–19–283039–2 (pbk.)

Printed in Great Britain by
Richard Clay (The Chaucer Press) Ltd
Bungay, Suffolk

TO JEAN

ACKNOWLEDGEMENTS

I SHOULD like to thank the original publishers of these essays for their permission to include them in this volume. The following list gives details of their relevant prior publication. Essays marked with an asterisk were previously collected in *The Art of the Soluble* (London, 1967: Methuen), those marked with a dagger in *The Hope of Progress* (London, 1972: Methuen).

'Introduction: Pluto's Republic' is newly written, but includes passages from the Introduction to *The Art of the Soluble*.

★ 'Two Conceptions of Science': 'Anglo-Saxon Attitudes', Henry Tizard Memorial Lecture, *Encounter* 143 (August 1965).

† 'Science and Literature': Romanes Lecture for 1968, *Encounter* 32 no. 1 (January 1969).

† 'Further Comments on Psychoanalysis': in *The Hope of Progress*.

'Induction and Intuition in Scientific Thought': Jayne Lectures for 1968 (Philadelphia, 1969: American Philosophical Society; London, 1969: Methuen).

★ 'Hypothesis and Imagination': *Times Literary Supplement*, 25 October 1963; repr. in expanded form in *The Art of the Soluble*.

'Victims of Psychiatry': review of I. S. Cooper, *The Victim is Always the Same*, *New York Review of Books*, 23 January 1975.

★ 'Darwin's Illness': review of Phyllis Greenacre, *The Quest for the Father*, and Gavin de Beer, *Charles Darwin*, *New Statesman*, 3 April 1964.

'Type A Behaviour and Your Heart': Introduction to Meyer Friedman and Ray H. Rosenman, 'Type A Behaviour and Your Heart' (London, 1974: Wildwood House).

'The Crab': review of Thelma Brumfield Dunn, *The Unseen Fight Against Cancer*, Larry Agran, *The Cancer Connection: And What We Can Do About It*, Priscilla Laws, *X-rays: More Harm Than Good?*, Lawrence LeShan, *You Can Fight For Your Life: Emotional Factors in the Causation of Cancer*, and Jane E. Brody and Arthur I. Holleb, *You Can Fight Cancer and Win*, *New York Review of Books*, 9 June 1977.

'Unnatural Science': review of Leon J. Kamin, *The Science and Politics of IQ*, and N. J. Block and Gerald Dworkin (eds), *The IQ Controversy*, *New York Review of Books*, 3 February 1977.

'Technology and Evolution': in *Technology and the Frontiers of Knowledge*, the Frank Nelson Doubleday Lectures, 1972–3 (New York, 1975: Doubleday).

'Does ethology throw any light on human behaviour?': in P. P. G. Bateson and R. A. Hinde (eds), *Growing Points in Ethology* (Cambridge, 1976: Cambridge University Press).

'Taking the Measure of Man': review of D. W. Forrest, *Francis Galton: The Life and Work of a Victorian Genius*, *Times Literary Supplement*, 24 January 1975.

* 'Herbert Spencer and the Law of General Evolution': 'Onwards from Spencer: Evolution and Evolutionism', Spencer Lecture for 1963, *Encounter* 120 (September 1963).

* 'D'Arcy Thompson and *Growth and Form*': 'Postscript: D'Arcy Thompson and *Growth and Form*', in Ruth D'Arcy Thompson, *D'Arcy Wentworth Thompson, The Scholar-Naturalist, 1860–1948* (London, 1958: Oxford University Press).

* '*The Phenomenon of Man*': review of Pierre Teilhard de Chardin, *The Phenomenon of Man*, *Mind* 70 (1961).

* '*The Act of Creation*': 'Koestler's Theory of the Creative Act', review of Arthur Koestler, *The Act of Creation*, *New Statesman*, 19 June 1964; Koestler's reply and P.B.M.'s further comments, ibid., 10 July 1964.

† 'J.B.S.': 'A Johnsonian Scientist', *New York Review of Books*, 10 October 1968.

† 'Lucky Jim': review of J. D. Watson, *The Double Helix*, *New York Review of Books*, 28 March 1968.

'House in Order': review of Barbara Ward and René Dubos, *Only One Earth*, and John Maddox, *The Doomsday Syndrome*, *World*, 12 September 1972.

* 'A Biological Retrospect': Presidential Address to Section D of the British Association, *Nature*, 25 September 1965.

'Expectation and Prediction' appears here for the first time.

† 'Science and the Sanctity of Life': *Encounter* 27 no. 6 (December 1966).

† 'On "The Effecting of All Things Possible"': Presidential Address to the British Association, 1969, in *The Hope Of Progress*.

I renew my thanks to Mr Arthur Koestler for allowing me to reprint his reply to my review of *The Act of Creation*.

My secretary and assistant Mrs Joy Heys has given me invaluable help in the preparation of this book for press, and I have received more editorial assistance from Dr Henry Hardy – an officer of the Press – than any author is entitled to expect.

P.B.M.

October 1981

CONTENTS

INTRODUCTION: PLUTO'S REPUBLIC

A GOOD many years ago a neighbour whose sex chivalry forbids me to disclose exclaimed upon learning of my interest in philosophy: 'Don't you just adore Pluto's *Republic*?'

Pluto's Republic has remained in my mind ever since as a superlatively apt description of that intellectual underworld which so many of the essays in this volume explore. We each populate Pluto's Republic according to our own prejudices: for me its most prominent citizens are IQ psychologists, and all psychotherapists who apply psychotherapy to the victims of organic diseases of the nervous system. With them I include all who share Dr David Cooper's sneering depreciation of the notion of *cure* (p. 69, note 1). Because we have to do here with prejudices I hasten to include among citizens of the Republic all thinkers – if that is not too far-fetched a description – who regard 'rhapsodic intellection' as a more than adequate substitute for the humdrum process of ratiocination that was thought good enough by Socrates, Descartes and Kant.

Other prominent citizens include all practitioners of 'scientism', especially those who apply what they mistakenly believe to be the methods of science to the investigation of matters upon which science has no bearing whatsoever, such as the element of moral instruction embodied in Restoration comedy (to choose the example I give below to illustrate the follies of gratuitous quantification – of the worship of what Ernst Gombrich calls *idola quantitatis*). Not less offensive than scientism is 'poetism', the application of literary standards of valuation to scientific theories, and in general the substitution of Cowley's 'painted scenes and pageants of the brain' for that which scientists try to make out by the *lumen siccum* of Francis Bacon.

High office in the Republic is held by mystical theologians, and in my lecture on 'Science and Literature' I cite George Campbell's interpretation of mystical theology: it is a Prose Offering to the Almighty which, where a living sacrifice would of necessity have been deprived of life, has been deprived instead of – sense. The writing of the admirable Campbell is so little known that I quote in full[1] the passage in which he makes his point.

I remember to have seen it somewhere remarked, that mankind being necessarily incapable of making a present of anything to God, have conceived, as a succedaneous expedient, the notion of destroying what should be offered to him, or at least of rendering it unfit for any other purpose. Something similar appears to have taken place in regard to the explanations of the divine nature and attributes, attempted by some theorists. On a subject so transcendent, if it be impossible to be sublime, it is easy to be unintelligible. And that the theme is naturally incomprehensible, they seem to have considered as a full apology for them in being perfectly absurd.

The present collection includes the whole of *The Art of the Soluble* (a title which may also be thought to need a word of explanation – see below), and the three lectures to the American Philosophical Society on scientific method that went to make up *Induction and Intuition in Scientific Thought* – too grand a title for so short a book, perhaps excused by being a faithful description of its subject-matter. It includes also a number of essays from a collection entitled *The Hope of Progress*, now (like *The Art of the Soluble*) out of print, and in addition a number of lectures or essays that have not hitherto appeared in book form.

'The Art of the Soluble' as a title came from my review of Arthur Koestler's *The Act of Creation*: 'No scientist is admired for failing in the attempt to solve problems that lie beyond his competence. The most he can hope for is the kindly contempt earned by the Utopian politician. If politics is the art of the possible, research is surely the art of the soluble. Both are immensely practical-minded affairs.'

One or two rather malicious people whose reading of that earlier book could have gone no farther than the title ('I never seem to get any time for reading nowadays') took it for granted that I was advocating the study of easy problems which would yield quick returns from the scientists' investment of time. What I meant, *of course*, was that the art of research was the art of making difficult problems soluble by devising

---

[1] George Campbell, *The Philosophy of Rhetoric* (London, 1776), chapter 6. The word 'succedaneous' had never before passed my lips when I enquired what it meant. It means 'of the nature of a substitute'; indeed, 'substitute' itself is an excellent *succedaneum* for the phrase 'succedaneous expedient'.

means of getting at them. Certainly good scientists study the most important problems they think they can solve. It is, after all, their professional business to solve problems, not merely to grapple with them. The spectacle of a scientist locked in combat with the forces of ignorance is not an inspiring one if, in the outcome, the scientist is routed. That is why so many of the most important biological problems have not yet appeared on the agenda of practical research.

So many passages in this book are critical of psychoanalysis and psychoanalytic theory that some defence is called for. To rebut a familiar imputation, let me in the first place say that I have never been injured by – have indeed never received or to my knowledge needed – psychoanalytic treatment, so that my critical attitude towards it has not the character of a revenge.

Many good people contend that, while the theoretical foundations of psychoanalysis leave much to be desired, yet, as a method of treatment, it demonstrably works: people do sometimes, perhaps often, get better as a result of psychoanalytic treatment – we have their own word for it, and who should know better?

That some people have improved *under* psychoanalytic treatment can hardly be a matter of dispute, but for reasons enlarged upon in the essay to which I refer below this is methodologically very poor evidence that they get better as a consequence of psychoanalytic treatment as such.

As to the theoretical foundations of psychoanalysis, in the chapter titled 'Further Comments on Psychoanalysis' (further, that is, to those made in 'Science and Literature') I raise a number of, as they seem to me, very damaging methodological criticisms of psychoanalysis that so far as I am aware have never been rebutted or explained away. Not all my criticisms are valid, however. It is often said, for example, that psychoanalysis achieves a complete intellectual closure by explaining away disbelief and indeed attributing disbelief to an infirmity of the psyche which only psychoanalysis can cure. Brian Farrell in his admirable new study *The Standing of Psychoanalysis*[1] has shown that this criticism is unjust, for although individual psychoanalysts may be injudicious enough to take this view it is not formally speaking any part of psychoanalytic theory. In just the same way many individual

[1] (Oxford, 1981).

Marxists believe that disbelief in Marxism reveals one's supine dependence on bourgeois ideology, but this does not follow directly from
Marxist theory and should not therefore be debited to it. In the
chapter to which I refer, I cite a whole number of psychoanalytic aetiologies and do not believe that any honest-minded man can read them
– especially that which relates to a matter so grave and socially so
important as anti-Semitism (p. 67) – without thinking that, so far from
being unjust, my reference to their 'Olympian glibness' was if anything
rather temperate. I single out for special censure (ibid.) the passage that
deals with the psychotherapy of ulcerative colitis, a disease now widely
regarded as 'auto-immune' in character – as a miscarriage of the immunological defence system in which the immunological defences of
the body have turned against one of its own constituents.

Worse still is the psychoanalytic treatment of *Dystonia musculorum
deformans* (DMD) which I came to hear of through Dr Irving S.
Cooper's[1] *The Victim is Always the Same*, of which 'Victims of Psychiatry' is a review.

Lovers of *Alice* will remember the scene in Looking-Glass World in
which the White King waits impatiently for the arrival of his two
messengers and asks Alice to look out for them. Having first seen
Nobody – a feat upon which the King compliments her, having regard
to the poor light – she sees one of them, Haigha (pronounced to rhyme
with mayor), bearing down on them very slowly. 'What curious attitudes he goes into,' says Alice, for the messenger keeps skipping up and
down, and wriggling like an eel as he comes along with his great hands
spread out like fans on each side. 'Not at all [curious],' says the King.
'He is an Anglo-Saxon messenger – and those are Anglo-Saxon attitudes. He only does them when he's happy.'

In reality poor Haigha was probably the victim of DMD, a dreadful
nervous disease of children and young adults marked by involuntary
muscular movements and grotesque postural fixations. It is extremely
unlikely that Haigha was happy: very probably he was deeply ex-
hausted and unhappy, for if DMD is incorrectly treated or not recognised as a physical disorder of the brain, it can destroy the victim's
personality and, in extreme cases, the victim's family too. Dr Cooper
(when Director of the Department of Neurosurgery in St Barnabas

---

[1] Author also of *Involuntary Movement Disorders* (New York, 1969).

Hospital, New York) devised a neuro-surgical treatment of DMD – a treatment that grew conceptually out of his spectacular successes with the neuro-surgical treatment of the uncontrollable tremors of Parkinsonism. The treatment of DMD depends upon the destruction of a nerve centre in the thalamus, a part of the brain that exercises an important controlling influence on the discharge of the nerve impulses which control muscular movement. The precise small area in the thalamus that must be destroyed cannot be known with certainty beforehand but is discovered during the course of the operation by direct observation of the improvement it brings about. To make this procedure possible the patient must be operated upon in the conscious state (which sounds horrific, but is not so because the brain does not feel pain). Sufferers from DMD who had the bad luck to fall into the hands of psychoanalysts are among the 'victims of psychiatry' to whom the essay so titled refers.

Darwin's illness is still a lively topic of discussion, partly because of the light it throws upon the degree to which non-clinical or non-scientific motives may sometimes prompt clinicians and scientists to take one view rather than another, and partly because of what it reveals about the thought-processes of psychoanalysts – the consideration that prompted me to study what they had to say about the matter. These criticisms have never been answered except by the innuendo that I was myself in need of psychoanalysis.

Leaving psychoanalytic interpretations on one side I recur to the theme that non-clinical and non-scientific considerations have played an enormously important part in the differential diagnosis of Darwin's illness. Some maintain that he did *not* have Chagas' disease because others maintain that he did, and I know of one expert on tropical medicine who emphatically denied the possibility in order to cut down to size a colleague who had received, as he himself had not, the high distinction of election into the Fellowship of the Royal Society of London. Certainly Darwin's physicians, with their comfortable diagnosis of hypochondria, have been defended out of a sense of professional solidarity against the actual or implied criticisms of mere scientists such as Saul Adler, Gavin de Beer, P. C. C. Garnham and myself.

Since I wrote my essay more has been learned of the therapeutic enormities of which poor Darwin was the victim. John H. Winslow[1] thinks it likely that Darwin suffered from chronic arsenical poisoning and perhaps also from mercury poisoning through the then common-place administration of mercurous compounds such as calomel. W. D. Foster,[2] studying Darwin's diary of his health, records the rather tragic entries that describe his responses to Dr Gully's 'water cure', of which a lurid description is given by Dr Lawrence A Kohn.[3]

Although they very probably made matters worse, I do not believe that Darwin's doctors killed him: that was the work, surely, of *Trypanosoma cruzi*. Sir George Pickering[4] opts for 'psycho-neurosis' – an interpretation with which I completely agree for the reasons I give in my essay, and which now seem to me to have gained strength from this newer evidence of the treatments Darwin suffered at the hands of his physicians.

I wrote above of the relevance of new evidence having to do with the pathology of Chagas' disease. A leading article in the *Lancet*, the principal British medical journal,[5] criticising by implication the widespread belief that damage to the musculature of the heart is the only or even the principal manifestation of Chagas' disease, summarises evidence which points to the occurrence of lesions in the stomach, duodenum, appendix, colon, gall-bladder, bile ducts, urinary bladder and ureter. These lesions give rise to a variety of symptoms which, with no aware-ness or reasonable presumption of their pathological connectedness, must have convinced Darwin's physicians of the rightness of their diagnosis of hypochondria – and have clearly misled the clinicians who have stood at Darwin's grave-side since. One of these clinicians, I confess, shocked me by producing evidence[6] of a kind which until then I had believed impressed only laymen: Darwin had persuaded some of his companions to be bitten by the 'great black bug of the pampas' which he out of curiosity had allowed to bite himself, but these others

[1] *Darwin's Victorian Malady: Evidence for its Medically Induced Origin* (Phila-delphia, 1971).

[2] 'A Contribution to the Problem of Darwin's Ill Health', *Bulletin of the History of Medicine*, **39**, 476–7, 1965.

[3] 'Charles Darwin's Chronic Ill Health', *Bulletin of the History of Medicine*, **37**, 239–55, 1963.

[4] *Creative Malady* (London, 1974).

[5] *Lancet*, 29 May 1965, 1150–1.

[6] Sir George Pickering, op. cit. (note 4 above).

– we were told – did not contract Chagas' disease. Such evidence is worthless, for if one of them *had* contracted Chagas' disease, how should we know? No such diagnosis would then have been possible. Besides, the symptoms of Chagas' disease are too irregular and fitful to justify the expectation of their taking the same shape in Darwin's companions as they took in him. Moreover, their state of health would not have been scrutinised as anxiously as Darwin's was.

A digression on mind and matter

The somewhat boisterous air of many of my criticisms of psychoanalysis has given rise to the impression that I believe soma and psyche are as distinct from each other as germ-plasm and soma – and, correspondingly, that I do not believe in psychosomatic influences or that states of mind can influence states of health. I can however defend myself from my own pages, for one essay ('Type A Behaviour and Your Heart') deals at length and sympathetically with the notion of Friedman and Rosenman that the state which conduces to coronary arterial disease is the consequence of the state of mind and the 'complex of personality traits' whose possessors they classify as 'Type A'.

I think their case is overstated – very few aetiologies are as simple and straightforward as this; moreover there are still some awkward gaps in the causal pedigree of coronary arterial disease. One of the world's leading epidemiologists once told me, though: 'We are all sure there is something in it but it is being very difficult to get decisive evidence.'

In the lengthy review ('The Crab') discussing a number of recent books on cancer written for general readers I pay specially close attention to Lawrence LeShan's views on the relevance of emotional factors to susceptibility to cancer, which LeShan believes to be often associated with that deep unhappiness which undermines one's sense of purpose and *raison d'être* – with the acquiescent passionless grief which Elizabeth Barrett Browning described in her famous sonnet, 'Grief':

> I tell you, hopeless grief is passionless;
> That only men incredulous of despair,
> Half-taught in anguish, through the midnight air
> Beat upward to God's throne in loud access
> Of shrieking and reproach. Full desertness
> In souls, as countries, lieth silent-bare

Under the blanching, vertical eye-glare
Of the absolute Heavens. Deep-hearted man, express
Grief for thy Dead in silence like to death:—
Most like a monumental statue set
In everlasting watch and moveless woe,
Till itself crumble to the dust beneath.
Touch it: the marble eyelids are not wet;
If it could weep, it could arise and go.

I mention in 'The Crab' the discovery that the outwardly visible vascular manifestations of the tuberculin reaction (the Mantoux test) can be abrogated by hypnotism – an observation of especial interest because of a certain family likeness between immunological resistance to tumours and to tubercle bacilli.

In spite of evidence of this general kind, there are still diehards among clinicians and among philosophers who do not believe that states of mind can influence states of health: that psyche and soma can indeed interact. The example I give to convince philosophers – it has convinced every philosopher to whom I have put it – is that of *blushing*. Blushing is caused by the closure of the arterio-venous anastomoses of the skin – of the blood vessels that transport blood directly from the smallest arterioles to the smallest venules, so bypassing the skin's surface. The closure of the arterio-venous anastomoses causes blood to flood into the capillaries of the skin, which reddens and feels warm. The reflex closure of arterio-venous anastomoses is surely a physical phenomenon, and no event could more plausibly be described as 'mental' than the recollection of a past embarrassment, yet such a recollection will sometimes cause people of sensibility to blush – as convincing an example of the action of mind upon matter as one could wish for.

The influence of states of mind upon states of health is not the form in which the mind–body problem most often presents itself to philosophers. For philosophers the problem is rather the nature of the connection, if any, between thinking, willing, feeling etc. on the one hand, and on the other hand the various physical performances of the nervous system, especially the busy traffic of nerve impulses.

It is a problem upon which two extreme views have been held: at one extreme, that mind is a thing apart and cannot be said to be in any way embodied – for mind belongs to a quite different 'semantic category' from nerve impulses and the like. At the other extreme is the uncompromising materialism that is embodied in Charles Darwin's question: 'Why is thought being a secretion of brain more wonderful than gravity a property of matter?'[1] Without going to the other extreme, as Darwin did, I feel confident that the dismissive 'category' argument is principally a defence by orthodox philosophers against what they have interpreted as another attempted usurpation of their subject-matter by those pesky scientists. It is a poor argument anyway: heredity and high-molecular-weight polymers also belong to different semantic categories; nevertheless genetic memory is physically embodied in the order of the nucleotides which, strung together, form the giant polymer deoxyribonucleic acid (DNA). I shall try now to explain the notion of a semantic category.

Consider a sentence such as 'The cat sat on the mat' and imagine a blank, to be filled in arbitrarily, in place of the word 'cat':

The . . . sat on the mat.

Clearly we could substitute the word 'dog', 'mouse', or – meaningfully, though implausibly and perhaps mistakenly – 'elephant'. On the other hand we could not substitute 'foreign exchange deficit', which would not be just erroneous or unlikely, but downright senseless because it belongs to a different semantic category. Some philosophers, led by Gilbert Ryle, have taken the view that thinking, willing and other such acts of mind belong to a different semantic category from nerve impulses and other such traffic of the brain. To attribute an act of mind to something that goes on in the brain is thus a category-blunder as elementary as to say that 'The case for proportional representation sat on the mat.'

After the publication of Gilbert Ryle's *The Concept of Mind* mention of categories and 'category-mistakes' became painfully common. I don't think, though, that many who used the term really understood what a 'semantic category-mistake' was, or that they would have been able to give tongue to whatever vague conception of it they may have had. Probably they took Ryle on trust, though to be sure the 'category-mistakes' to which Ryle refers are in reality simply mistakes – quite

[1] I learned of this passage in one of Darwin's transmutation notebooks from Stephen J. Gould's admirable essays *Ever Since Darwin: Reflections in Natural History* (New York, 1977).

straightforward and easily understandable mistakes, too, such as any-one might make.

One example of what Ryle calls a category-mistake comes to mind: he envisages a foreigner in Oxford who is shown a number of colleges, libraries, playing fields, museums, scientific departments and admini-strative offices, and then asks where the University is – thus making the elementary category-blunder of confusing an abstract pedagogic entity, the University, with a piece of ground occupied by various kinds of academic masonry. But this is not a semantic category-mistake, it is simply a mistake – one that might easily be made by Americans used to the idea that universities are real material objects situated on a campus, as so many American universities are. Popper has criticised Ryle's argument in detail elsewhere[1] and I agree with him that there is nothing in the concept of semantic category-mistakes which prohibits our thinking that states or acts of mind can exercise physical effects.

*

In an earlier paragraph I classified IQ psychologists among the more prominent citizens of Pluto's Republic: 'Unnatural Science' is the article that investigates their pretensions. I still think that the distinc-tion between natural and unnatural sciences is a useful one but because of misunderstandings I hasten to add that I do *not* count sociology among the latter. I do not think that sociology's being described as a 'behavioural science' has done it a bit of good: it is as if some special characteristic made it necessary to relegate sociology to a sort of ghetto of learning. The best sociology is simply a science *tout court* of which the subject-matter happens to be society and human social behaviour. 'Unnatural Science' pays special attention to two special cultural aber-rations embodied in IQ psychology: the belief that intelligence is amenable to a single-value scalar measurement, and the belief that it is possible to attach percentage values to the respective contributions of nature and nurture to any character difference that is influenced by both.

I chose to discredit the follies of quantification by examples drawn from demography, soil science and economics. But I could equally well have chosen modern computerised literary criticism if I had known

[1] See *Conjectures and Refutations* (London, 1963), chapter 13.

then Professor Harriett Hawkins's dazzling review[1] of a book purporting to show that Restoration comedy is 'a body of literature declaring moral instruction to be its main purpose'. A quotation from her review will illustrate the quality of thought underlying this remarkable conclusion:

It is rather fun to know that 'of the 37 whores' who appear in 22 of the plays here surveyed, 13 are rewarded with settlements or husbands . . . Almost as many whores as are rewarded are held up to contempt, but on close inspection it appears that whoredom is not the cause of their being shamed: 6 of these 11 exhibit ill humor when they do not get their way, a breach of decorum that the plays always punish; 3 others help cheats who must be exposed; and the remaining 2 are the only straightforward prostitutes who are disapproved of.

Similarly quotable statistics are scattered throughout Mr Schneider's study: 'Of 16 young men who marry for money in the plays here under consideration 15 are poor.'

In 'Unnatural Science' a good deal turns on the special point that culturally mediated inheritance (exogenetic heredity) awards an enormous evolutionary premium to those who are teachable and able to teach. Exogenetic heredity is discussed in the lectures 'Technology and Evolution' and 'Does ethology throw any light on human behaviour?' The existence in human beings of a system of heredity not mediated through the chromosomes was first recognised by the leading geneticist of his day: Thomas Hunt Morgan.[2]

It is hard to leave the subject of IQs without mention of Sir Francis Galton, my essay upon whom ('Taking the Measure of Man') refers to the most disagreeable trait of IQ psychologists. I have in mind Galton's almost exultant contempt for people who try to aim higher in life than their so-called 'innate' intelligence is thought to authorise. Many students of the matter will have noticed the recurrence of a similar train of thought in the writings of modern advocates of IQ psychology.

Many IQ psychologists do not appear to me to be very bright by the standards that prevail widely among educated people. In particular, some IQ psychologists are evidently completely unable to understand

---

[1] *Review of English Studies*, **23**, 209–11, 1972.
[2] Thomas Hunt Morgan, *The Scientific Basis of Evolution* (London, 1932), chapter 10.

the argument – spelt out in full in 'Unnatural Science' – which demonstrates the wrong-headedness of attempting to attach exact percentages to the respective contributions of nature and nurture to character differences. Nor is this the only example of straightforward up and down logical reasoning that they seem not to be able to grasp. Another is the element of bias – of logical asymmetry – in the argument that IQ scores must signify something deeply important because of the positive correlation between high IQ scores in childhood and success in later life (measured, characteristically, by income); but there is here a grave bias of ascertainment. We all know of people who did well in life, even in scholarly pursuits, in spite of having got low IQ scores or low marks in exams based on the IQ ideology. These damaging anomalies come to light because the success of the subjects provides an incentive to enquire into their scholastic background; there is no comparable incentive for finding out about people who did badly in life in spite of having had high IQ scores: *Die im Dunkeln sieht Man nicht*, Brecht observes – those in darkness drop from sight. There is also the very disturbing self-fulfilling element in expectations or predictions based upon IQ scores or something comparable to them. In a society politically committed to a belief in the predictive value of these scholastic performances in childhood the low scorer will almost automatically be denied scholastic preferment, and also therefore whatever contribution this makes to success in life, for in the spirit of Francis Galton IQ psychologists would judge it fruitless to spend time and money on persons who have scored low in their exercises. In Pluto's Republic it is taken for granted that intelligence can be measured by procedures that are free from cultural bias and are uninfluenced by acquired skills in the use of language and so forth. When a person is 5 feet 10 inches this is referred to as his or her height, not as a measurement of his height, and the IQ boys have tried to convince us that a person's IQ *is* his intelligence, which is on all fours with saying that a patient's temperature is his health or that the time he takes to run a mile is his degree of athleticism.

In so far as the essays collected here have a central or recurrent theme, it is in the attempt to answer the following questions: What is science, what kind of person is a scientist, and what kind of act of reasoning leads to scientific discovery and the enlargement of the under-

standing? I seldom put these questions directly, and answer them only
incompletely and bit by bit; but, taking the first question first, it
seemed natural to put 'Two Conceptions of Science' first in this collection. This was the Henry Tizard[1] Memorial Lecture for 1965. The two
conceptions are, roughly speaking, the romantic and the rational, or
the 'poetic' – in Shelley's sense ('poetry comprehends all science') –
and the analytical, the one speaking for imaginative insight and the
other for the evidence of the senses; one finding in scientific research
its own reward, the other calling for a valuation in the currency of
practical use. None of these distinctions is really satisfactory, for the
sets of opinions do not hang together logically or rationally in themselves; rather they are complexes of opinion that tend to go with
certain temperaments, much as Tory and Labour or Republican and
Democrat stand for casts of thought as much as for casts of vote.
Nevertheless I think I was right in saying that in the romantic conception

truth takes shape in the mind of the observer: it is his imaginative preconception
of *what might be true* that provides the incentive for finding out, so far as he can,
what *is* true. Every advance in science is therefore the outcome of a speculative
adventure, an excursion into the unknown. According to the opposite view,
truth resides in nature and is to be got at only through the evidence of the
senses: apprehension leads by a direct pathway to comprehension, and the
scientist's task is essentially one of *discernment*.

Thomas Henry Huxley was a firm believer in the idea that the task
of the scientist was essentially to imbibe the information voluntarily
proffered by nature; Huxley translated a German nature poem for the
first issue of the scientific magazine *Nature*, the purpose of which, he
said, was 'to mirror the progress of that fashioning by Nature of a
picture of herself, in the mind of Man, which we call the progress of
science'.[2]

I am not now at all satisfied with the argument that occupies the

[1] In his day Sir Henry Tizard (1885–1959) was England's foremost scientific
administrator, and he was the first scientist in recent times to become the Head of
an Oxford College. Tizard played an important part in winning the war by his
enormously energetic advocacy of the development and installation of radar. He
also devised the method of rating the detonative properties of gasoline by what
later came to be called the 'octane number'.

[2] See W. H. Brock, *Nature and Victorian Imagination* (Berkeley, 1978).

latter half of my lecture, which deals with why, in the English-speaking world especially, a class distinction has grown up around the difference between 'pure' and 'applied' science. My suggestion ran as follows: in the English-speaking world, poetic invention is the paradigm of pure creative activity. The Romantics shunned 'applied' poetry – poetry for the occasion, or poetry upon a given theme – because they felt that artistic creation should be a natural and spontaneous upwelling of ideas. Correspondingly, the highest form of science must be that which is spontaneously proffered by the creative imagination, not something wrung from us by the pressure of necessity.

This interpretation struck me as plausible not only because the Romantics championed it, but also because our reverence for pure science in the modern sense (see below) does in fact date from the first half of the nineteenth century. In the seventeenth and eighteenth centuries the idea of science pursued for its own sake was regarded as frivolous or even comic: that is why Francis Bacon and Thomas Sprat had to beg for interpretative research, for *experiments of light*. Sir Nicholas Gimcrack, Thomas Shadwell's *Virtuoso*, full of the 'purest' scientific enterprises, is an egregious idiot. Goldsmith had to plead the usefulness of his *History of the Earth and Animated Nature* because the opposite of useful was not pure but – *idle*. Swift's ridiculous Laputans were scientists – especially Fellows of the Royal Society. It went with this cast of mind that the devising of hypotheses was judged a perilous indulgence; the 'Creative Imagination' was looked upon with disfavour and the word 'enthusiasm' had a pejorative flavour.

But Coleridge was not against the Useful Arts. Agriculture, commerce and manufactures, he told us,[1] 'are now considered scientifically', and are indeed founded on the sciences: 'It is not, surely in the country of ARKWRIGHT, that the Philosophy of Commerce can be thought independent of Mechanics: and where DAVY has delivered Lectures on Agriculture, it would be folly to say that the most Philosophic views of Chemistry were not conducive to the making our valleys laugh with corn.'

Coleridge certainly thought the Useful Arts less noble than the Pure Sciences – but because of their fallibility, not because of their usefulness. Coleridge used 'pure' in its older and original meaning. The Pure

[1] Coleridge, *On Method*, 3rd ed. (London, 1849).

Sciences are those of which the axioms are derived, not from experience, but by intuition or revelation: the 'Pure Sciences . . . represent pure acts of the Mind', Coleridge says; '. . . at the head of all Pure Science stands *Theology*, of which the great fountain is Revelation'. Pure Sciences are best not because they lack practical application but because the revealed or intuitively certain axioms out of which they grow are not subject to error.

Over the past hundred years or so the word 'pure' in this context has undergone a slow revolution of meaning which has allowed its lesser connotation to usurp the greater. 'Pure science' now means that which is done without regard to or interest in application, use or material gain, and we speak of 'pure physics' or 'pure biological research' in a sense that differs in an important way from Coleridge's. Yet the idea of the superiority of pure over applied science has remained, and with it the dire equation *Useless = Good*.

Many other influences were of course at work. Perhaps the most important of all – and the best understood, in spite of my failure to have given it due weight – was the very deliberate perpetuation by English 'public schools' (private secondary boarding schools) of the Aristotelian conception of activities that did or did not become a gentleman.[1] It was admirably well put by John Gillies in the introduction to his translation of Aristotle's *Ethics and Politics*.[2] Aristotle's experimentations, he tells us, were 'confined to catching Nature in the fact, without attempting, after the modern fashion, to put her to the torture'; for philosophers

ranked with the first class of citizens; and, as such, were not lightly to be subjected to unwholesome or disgusting employments. To bend over a furnace, inhaling noxious steams; to torture animals, or to touch dead bodies, appeared to them operations . . . unsuitable to their dignity. For such discoveries as the heating and mixing of bodies offer to inquisitive curiosity, the naturalists of Greece trusted to slaves and mercenary mechanics, whose poverty or avarice tempted them to work in metals or minerals . . . The work-shops of tradesmen then revealed those mysteries which are now sought for in colleges and laboratories.

[1] See J. C. Dancy's Presidential Address to the Educational Section of the British Association in 1965, in *The Advancement of Science*, **22**, no. 100, 379, 1965.
[2] (London, 1797), vol. I, pp. 139–40.

These canons of gentility were still in force and gloried in during my own schooldays. They are quite largely responsible for the position Great Britain holds in the world today. It is not envy or malice, as so many people think, but utter despair that has persuaded so many educational reformers to recommend the abolition of the English public schools.

I spoke of only *two* conceptions of science and make amends now by calling attention to the most popular of all, a third: it is that which distinguishes the old-fashioned humanist and the modern literary intellectual. A scientist is a man of facts and calculations, empty of all poetry and all wonder, who operates a calculus of discovery and arrives at the truth by what is essentially a mechanical procedure. This was how Blake saw him, and Keats and Shelley, but – characteristically – not Coleridge; it was how Edgar Allan Poe saw him too, for whom science was 'a vulture whose wings are dull realities'.

My wife and co-author Jean lighted upon some passages in Dickens's *Hard Times* that describe Inductive Man better than ever I could. They represent the opinions of a pedagogue, Thomas Gradgrind. And who, pray, was he?

'Thomas Gradgrind, sir. A man of realities. A man of facts and calculations. A man who proceeds upon the principle that two and two are four, and nothing over, and who is not to be talked into allowing for anything over. Thomas Gradgrind, sir – peremptorily Thomas – Thomas Gradgrind. With a rule and a pair of scales, and the multiplication table always in his pocket, sir, ready to weigh and measure any parcel of human nature, and tell you exactly what it comes to. It is a mere question of figures, a case of simple arithmetic. You might hope to get some other nonsensical belief into the head of George Gradgrind, or Augustus Gradgrind, or John Gradgrind, or Joseph Gradgrind (all suppositious, non-existent persons), but into the head of Thomas Gradgrind – no, sir!'

Seeing wonder as the enemy of ratiocination and hearing his daughter and pupil Louisa one day saying to her brother: 'Tom, I wonder . . .' Mr Gradgrind stepped forward and said 'Louisa, never wonder!'

Herein lay the spring of the mechanical art and mystery of educating the reason without stooping to the cultivation of the sentiments and affections. Never wonder. By means of addition, subtraction, multiplication, and division, settle everything somehow, and never wonder. Bring to me . . . yonder baby just able to walk, and I will engage that it shall never wonder.

Rationalist that he was, the hopeful and benignant Marquis de Condorcet was inductivist enough to believe that science had no longer anything to guess: 'All it has to do is to collect and arrange facts and exhibit the useful truths which arise from them as a whole and from the different bearings of their several parts.'

After 'Two Conceptions of Science' I have put 'Science and Literature', a Romanes Lecture delivered in Oxford University, which was taken by at least one literary critic to be a rather wicked attack upon literature and the humane values that go with it. It is in reality an attack upon two widely prevalent cultural diseases, *scientism* and *poetism*. Of scientism I have given in earlier paragraphs a splendid example borrowed from Professor Harriett Hawkins; scientism has other powerful opponents too.[1]

But what about poetism, the hero of which is that William Blake who 'came in the grandeur of inspiration to cast off rational demonstration'?[2]

We are still so deeply steeped in the tradition of which Blake was one of the founders that we find it difficult to realise that the intensity of the conviction with which we believe a theory to be true has no bearing upon its validity except in so far as it produces a proportionately strong inducement to find out whether it is true or not. Some theories have enjoyed an unduly long lease of life because of a certain dark visceral appeal. I remember one such theory from my student days. It was at one time believed that the salty composition of the blood of human beings and other land vertebrates faithfully reproduced the salty composition of the oceans in the era when vertebrate animals were beginning to leave the sea to colonise dry land. The idea that our blood represents an entrapment into the body of some dark Silurian sea gratifies that taste for Gothic mystery which in one form or another is present in us all. But then – and here indeed is Blake's Spectre of Rational Thought come to 'cast a mildew' over the whole inspired conception – there are no grounds at all for believing that the blood or

[1] See for example Jacquetta Hawkes's Danz Lecture, *Nothing But or Something More* (Washington, 1972).
[2] Blake, *Milton*, book 2.

other body fluids of vertebrate animals were ever in that osmotic and ionic equilibrium with sea-water which this theory requires.

Friedrich Nietzsche once contemptuously dismissed an opinion which struck him as culpably idiotic as 'an inscription over the porch of a modern lunatic asylum'. One criticism of my Romanes Lecture – in the *Times Literary Supplement* of 23 June 1972 – equally deserves to be preserved in an engraving surrounded by ornamental scrollwork: 'The real nature of the challenge to science . . . is a claim that the essential method of scientific thought inevitably leads to a rejection of much that is of most value in human experience, the denial of values, of beauty, of humane qualities in general.' If I made no mention of any such opinion it is because I have never held so contemptuous a view of my colleagues in the humanities as to suppose them capable of holding it.

'Science and Literature' was quite a suitable topic for my contribution to the series of lectures founded by George John Romanes (1848–94), because in addition to being a versatile scientist Romanes was reputed to be a poet. A single quotation from his writing will show that his reputation is not well founded. We are to imagine Romanes in a paroxysm of grief at the sheer unreasonableness of Darwin's being taken from us. 'The struggle cease,' he counsels himself:

> And when the calm of Reason comes to thee,
> Behold in quietness of sorrow peace.
> By such clear light e'en in thine anguish see
> That Nature, like thyself, is rational;
> And let that sight to thee such sweetness bring
> As all that now is left of sweetness shall:
> So let they voice in tune with Nature sing,
> And in the ravings of thy grief be not
> Upon her lighted face thyself a blot.

This poem was specially selected for us by a distinguished man of letters and a former President of Magdalen College, T. Herbert Warren; but *why*, God only wot.

'Science and Literature' is followed by the 'Further Comments on Psychoanalysis' elicited by the critical response which the Romanes Lecture received.

The next few essays in the collection are an exposition of Karl

Popper's methodology of science. The three 'Jayne' Lectures to the American Philosophical Society entitled 'Induction and Intuition in Scientific Thought' are intended to be read consecutively to form a tiny textbook expounding the shortcomings of inductivism and the character of the 'hypothetico-deductive' scheme of thought which we associate with the name of Karl Popper. The essay that follows, 'Hypothesis and Imagination', is a history of the ideas underlying the Popperian scheme of thought, which goes back farther than many people realise. As this latter essay embodies a crash course in Popperism there are some overlaps between it and the Jayne Lectures. Popperian ideas pop up elsewhere, too. Inevitably there are overlaps,[1] then, and these are sure to annoy those who demand full value for money from every printed page. I regard these overlaps as a measure of the pervasiveness of Popper's ideas and of their high explanatory value in any context having to do with creativity and the functions of criticism.

Because of its practical-mindedness – because it deals with thought at bench or shop-floor level – Popper's methodology is coming to be accepted very widely as doctrinal by scientists and medical men. I must record, however, that people who work in the historical or literary traditions of thought tend to incline rather more to Dr Thomas S. Kuhn's opinions of the role of 'paradigms' in day-to-day science and in revolutions of thought.

Considered as a history of ideas 'Hypothesis and Imagination' is too much concerned with English and Scottish philosophical thought, and with other philosophic traditions only in so far as they touch upon or mingle with it. I now make amends for this confinement of my history by pointing out that Immanuel Kant had a pretty clear conception of many of the essentials of the hypothetico-deductive method, as we know from lectures published posthumously in English under the title *Kant's Introduction to Logic*.[2]

[1] And not only in my references to hypothetico-deductive thought. In a book composed of essays written over many years, it is inevitable that there should be overlaps of a kind that would have been avoided if the book had been conceived as a whole. I have excised the worst of these, but left the remainder as harmless or even rhetorically useful.
[2] Translated by T. K. Abbott, with notes by S. T. Coleridge (New York, 1963).

In my essay on Popper's philosophy I called special attention to the importance of Claude Bernard's excellent anticipation of the main outlines of the hypothetico-deductive scheme of thought. Writing in the *British Journal for the History and Philosophy of Science* a 'philosopher of science' once wrote with the utmost contempt of my trying to pass off Claude Bernard as an important philosopher of science. I therefore take this opportunity to reaffirm my opinion that he is so – a point upon which Popper, commenting upon my essay in his honour, agreed with me warmly.

Several essays in this collection deal with the work or thought of particular people. The Spencer Lecture begins with an account of the logical weakness of his or any other theory of general evolution – of any theory which declares that evolution is the fundamental stratagem of nature, is 'a general condition to which . . . all systems must bow . . . a light illuminating all facts, a curve that all lines must follow', to put the matter in words (Teilhard's) which would not have commended themselves to Spencer himself. Yet the theory of evolution in its ordinary technical sense is still one of the liveliest topics in biology, and the next great advance we may look forward to is a new theory of genetic variation, propounded by molecular biologists. For the reasons explained in my Presidential Address to the Zoological Section of the British Association ('A Biological Retrospect'), and in greater detail in 'Expectation and Prediction', it is impossible to predict what this advance will be, but I guess it will be something in the nature of a sentential calculus of genetic messages – a natural development of the glossary that is now being compiled of single words.

The latter part of the Spencer Lecture deals with certain abstract formal similarities between natural transformations of energy, information theory, probability and the idea of order; or, to put it in another way, between entropy, randomness and nonsense.[1]

Another lecture that deals with the work of a particular man is 'D'Arcy Thompson and *Growth and Form*'. Because D'Arcy Thompson was a classicist and a mathematician as well as a natural historian, it may seem specially significant that he believed, as I also believe:

[1] A humanist who has read as far into my book as this should at this stage be most earnestly exhorted to browse in *Home Life with Spencer by Two* [*sic*] (Bristol, 1906).

not merely that the physical sciences and mathematics offer us the only pathway that leads to a deep understanding of animate nature, but also that the true beauty of nature will be revealed only when we have worked our way to a deep understanding of natural dispositions. To us nowadays it seems obvious that the picture we form in our minds of nature will be the more beautiful for being brightly lit. To many of D'Arcy's contemporaries it must have seemed strange or even perverse that he should have combined a physico-mathematical analysis of nature with, at all times, a most intense consciousness of its wonder and beauty; for at that time there still persisted the superstition that what is beautiful and moving in nature is its mystery and its unrevealed designs. D'Arcy did away for all time with this Gothic nonsense: a clear bright light shines about the pages of *Growth and Form*, a most resolute determination to unmake mysteries.

This last sentence is my cue to mention my highly critical review of Pierre Teilhard de Chardin's *The Phenomenon of Man*. A good deal of Teilhard is nonsense, but on further reflection I can see it as a dotty, euphoristic kind of nonsense, very greatly preferable to solemn long-faced Germanic nonsense. There is no real harm in it. But what, I wonder, was the origin of the philosophically self-destructive belief that obscurity makes a prima-facie case for profundity? – the origin, I mean, of the comically fallacious syllogism that runs *Profound reasoning is difficult to understand; this work is difficult to understand; therefore this work is profound*.

In the seventeenth and eighteenth centuries philosophic writing was marked by its clarity. John Locke is much easier to follow than John Donne. Descartes had to assure even his most diffident readers[1] that there was nothing in his writings 'which they are not capable of understanding completely, if they took the trouble to examine them', and the ideas of clarity and of distinctness of vision occur as often in his writings as the idea of light in Bacon's or, indeed, in the writing of Comenius, as the title of his great work *Via lucis* reminds us.

David Hume and George Berkeley, Thomas Reid and Dugald Stewart, had definite opinions to express and took great care to make them fully understood; but with Kant things changed. Kant was a very profound philosopher, in some ways the greatest there has ever been. One of the reasons why he is hard to understand is that the problems he

---

[1] In a letter to the translator (into French) of his *Principles of Philosophy*.

tried to solve were intrinsically of the utmost complexity. We marvel at and revere his struggle for clarity. Unfortunately, his struggle was sometimes unsuccessful. Kant became notorious for his obscurity. Dugald Stewart, wrestling with the sage in Latin translation, spoke of his 'utter inability to unriddle the author's meaning . . . I have always been forced to abandon the undertaking in despair'. Seen through Peacock's eyes, the young Shelley delighted in 'the sublime Kant, who delivers his oracles in language none but the initiated can comprehend', and Coleridge 'plunged into the central opacity of Kantian metaphysics, and lay *perdu* for several years in transcendental darkness'. Kant was just the man for the Romantics.

The harm Kant unwittingly did to philosophy was to make obscurity seem respectable. From Kant on, any petty metaphysician might hope to be given credit for profundity if what he said was almost impossible to follow.[1] There grew up a new style of philosophic writing of which F. H. Bradley was the greatest English master. It seems to have affected many of its readers like a drug, and the intense resentment aroused by the work of linguistic philosophers might be thought of as part of a withdrawal syndrome, for in England metaphysics in the Bradleian style has almost completely disappeared. Of course we all need to be *bunkrapt* from time to time (Paul Jennings's word), but not by works that profess to be philosophy. 'I love to lose myself in a mystery,' wrote Thomas Browne. Scientists do not.

In retrospect I think my essay on Teilhard was good of its kind, but I confess that when on the insistence of an American writer friend I read Mark Twain's 'Fenimore Cooper's Literary Offences' I bowed my head in the presence of a master of literary criticism.

I do not budge from my description of Teilhard's work as so much 'obscure pious rant', but with hindsight I do think that I was coarsely insensitive in not reading Teilhard's work – or rather in not interpreting its great popularity – as a symptom of hunger, a hunger for answers to questions of the kind that science does not profess to be able to answer (questions that are loftily dismissed by positivists as non-questions or pseudo-questions). I wonder how long it will be before

---

[1] Kant had to face this charge himself: 'I am often accused of obscurity, perhaps even of deliberate vagueness in my philosophical discourse to give it the air of deep insight' (Preface, *The Metaphysic of Morals* (1797)).

positivists formally concede that inferences having to do with the first and last things cannot be deduced from propositions containing only empirical furniture – which have to do only with shoes and ships and sealing-wax and cabbages and kings.

One allusion in my Teilhard review caused some puzzlement and I know it exasperated translators into other languages. In one passage in *The Phenomenon of Man* Teilhard got himself into a metaphysical muddle from which it seemed impossible that he should be able to extricate himself; but by a piece of philosophic legerdemain he got out of it. 'With one bound', I remarked, 'Jack was free' – or as my bemused Spanish readers learned, *Con un salto Jack queda libre*.

I first read about Jack in George Orwell – I can't remember exactly where. At one time Orwell was making a living writing weekly instalments of an adventure serial for a boys' weekly. The formula to which such serials were written, like that of the old silent movie serials, was that each instalment ended with the hero(ine) in peril so extreme that next week's instalment just had to be bought to see how (s)he got out of it. When writing one such instalment Orwell got Jack into a situation so perilous that it seemed impossible he should be able to extricate himself. Orwell struggled at length and in detail with the beginning of the following week's episode until a colleague of his – a Fleet Street tradition says the office boy – took the copy from him, crossed it out and wrote instead 'With one bound Jack was free.'

My aged mother was very shocked by my review of Teilhard: 'How *could* you be so horrid to that nice old man?' she asked me. The reason, I told her, was that Teilhard had described his book as a work of science – and one executed with 'remorseless logic' – and as a work of science it has been accepted by its more gullible readers. If only he had described it as an imaginative rhapsody 'based on science' in much the same way as some films are said to be based on books to which in the outcome they seem to bear little resemblance, then *The Phenomenon of Man* would have caused no offence.

Most of the other pieces that deal with the thought or work of particular people – 'J.B.S.', 'Lucky Jim [Watson]' and 'Taking the Measure of Man' (about Galton) – are self-explanatory, but I ought to explain that 'House in Order' was written very consciously as an act of homage to the late Barbara Ward and to acclaim and marvel at her ability to grasp the gist or general sense of scientific ideas even though she was in no sense a scientist.

I did not expressly include futurologists among the citizenry of Pluto's Republic: they are recent immigrants, after all, but they are naturalised now and I very much hope that my essay 'Expectation and Prediction' does something deeply to undermine the pretensions of their weird farrago of scholarly pretensions and barefaced guesswork.

Throughout this book I defend science against the misrepresentations of mystical theologians and literary intellectuals. My defence of science does not take the form of making extravagant claims for the efficacy of scientific method in solving the many grievous problems that confront us as human beings. On the contrary: in 'Science and the Sanctity of Life' I put what I conceive to be the role of scientific evidence in the following way:

If the termination of a pregnancy is now in question, scientific evidence may tell us that the chances of a defective birth are 100 per cent, 50 per cent, 25 per cent, or perhaps unascertainable. The evidence is highly relevant to the decision, but the decision itself is not a scientific one, and I see no reason why scientists as such should be specially well qualified to make it. The contribution of science is to have enlarged beyond all former bounds the evidence we must take account of before forming our opinions. Today's opinions may not be the same as yesterday's, because they are based on fuller or better evidence. We should quite often have occasion to say 'I used to think that once, but now I have come to hold a rather different opinion.' People who never say as much are either ineffectual or dangerous.

I thought this a pretty modest statement and think so still. It fell far short of satisfying another participant in the congress at which this speech was delivered, a Mr St John-Stevas. He chose after my speech to express his complete rejection of the opinions of 'Professor Medawar', pronouncing the title with a highly pejorative inflexion presumably in the hope (not fulfilled) of currying favour with the audience. Mr Stevas's own contribution[1] gives him a permanent place in the dialogues of Pluto. It ended as follows:

The following story, which was told by Maurice Baring, is of some relevance. One doctor to another: 'About the terminating of pregnancy, I want your

[1] In D. H. Labby (ed.), *Life or Death: Ethics and Options* (Seattle, 1968).

opinion. The father was syphilitic. The mother tuberculous. Of the four children born, the first was blind, the second died, the third was deaf and dumb, the fourth was also tuberculous. What would you have done?' 'I would have ended the pregnancy.' 'Then you would have murdered Beethoven.'

This rabble-rousing utterance embodies a very grievous fallacy. Not even Mr Stevas can believe that a tuberculous mother and a syphilitic father are *more* likely than normal people to give birth to a musical genius. But if they are not, the whole story is pointless because the world is just as likely to be deprived of a Beethoven by chaste absten-tion from intercourse or by a woman's having a menstrual period.

I have put last in this book the lecture that I myself liked best both to prepare and to deliver: it was 'On "The Effecting of All Things Possible" ' – Francis Bacon's great Atlantidean dream of the purpose of science. I must add, though, that he puts the idea much more reasonably and clearly in his little-known *Valerius Terminus*, in which he says 'The true aim of science' is 'the discovery of all operations and all possibilities of operations from immortality (if it were possible) to the meanest mechanical practice'.

I think many people missed the point of my title. The whole history of science and of technology based on science bears witness that every-thing which is in principle possible will be done if the intention to do it is sufficiently resolute and long-sustained. If this is the great glory of science it also embodies its greatest threat, but (as my wife and I put it)[1] it makes no sense on the one hand to cower at the havoc science and technology may cause in achieving the Baconian ambition of effecting all things possible and at the same time to exclude from all things possible the discovery of remedies for technological malefactions.

In this final essay I compare the fearfulness and despondency that is characteristic of intellectuals – particularly literary intellectuals – today with the 'metaphysical gloom', helplessness, real or more often affected melancholy and (as Christopher Hill put it) 'loss of nerve' that was prevalent in the early seventeenth century, and have been surprised since to learn how many people seem not to realise that the darker Shakespeare of *Hamlet*, *Timon*, *Macbeth* and *Lear* was a Jacobean, not an Elizabethan dramatist. In any event, I thought the parallel with our

[1] *The Life Science* (London, 1977), chapter 24.

fashionable present-day distrust of science and even of logical reasoning was close enough to justify a defence such as I undertook.

David Wilson, Science Correspondent of the BBC, told me that my address reminded him of the kind of pep talk he sometimes had to put up with from his headmaster. I acknowledge the force of this criticism but plead in my defence that I was fulfilling a recognised function of the President of the British Association: to speak *for* science as well as *of* science. I felt it important to try to restore the faith of my colleagues in the notion of progress. 'Ah, but what do you mean by progress?' people say to each other, exchanging wise smiles and knowing looks. I learnt from Karl Popper not to put much trust in mullings over the 'true meanings' of words which are already in common use and intelligible to us all. Simple-minded people often discuss the word 'progress' as if it had some deep true inner meaning which discussion and sematological research would ultimately reveal.

Raymond Mortimer defined a 'highbrow' as a man who prefers good books to bad; and reassuming the role of the headmaster of David Wilson's parallel I feel I should now snap out: 'Hands up any boy who prefers bad drains to good drains, and leave the room for not believing in progress.' 'Ah, but that's just material progress', plaintive voices chip in to say, forgetting that some measure of material progress is necessary if the human spirit is fully to unfold, and heedless also of Brecht's *Erst kommt das Fressen, dann kommt die Moral*. Ill-disposed critics have spoken of my equating all progress to material progress – a travesty of my opinions. When I speak of our endeavours to make the world a better place to live in I neither say nor imply that this melioration can be achieved by purely material means, though I am quite sure it can't be achieved without them. That 'Material Progress implies Spiritual Impoverishment' is yet another Nietzschean inscription of the kind I referred to above – if not over the porch of a modern lunatic asylum, then, to bring out its nonsensicality, let us say over the gates of fifteenth-century Florence or Rembrandt's Amsterdam. I believe, as many others do, that material progress is necessary for our improvement, but I do not know, and have never heard of, anybody who said that it could be sufficient.

In the seventeenth century there was reason enough for melancholy. It was a settled conviction that the world had almost run out of its ration of time and might well therefore come to an end even within the lifetime of people then living. These were reasons enough for despondency, but there are no reasons of comparable gravity today. Today despair of the future and acquiescence in the notion that we are going steadily downhill are a shameful admission of a failure of nerve: for, as I declared in my lecture, 'to deride the hope of progress is the ultimate fatuity – the last word in poverty of spirit and meanness of mind'.

Nowadays most people are unwilling to contemplate – let alone to entertain – any set of beliefs that has no name attached to it. The general tone of my Presidential Address to the British Association may give the impression that I am an 'optimist', but indeed I am no such thing, though I admit to a sanguine temperament. I prefer to describe myself as a 'meliorist' – one who believes that the world can be improved by finding out what is wrong with it and then taking steps to put it right.

TWO CONCEPTIONS OF SCIENCE

My theme is popular misconceptions of scientific thought. I shall argue that the ideas of the educated lay public on the nature of scientific enquiry and the intellectual character of those who carry it out are in a state of dignified, yet utter, confusion. Most of these misconceptions are harmless enough, but some are mischievous, and all help to estrange the sciences from the humanities and the so-called 'pure' sciences from the applied.

Let me begin with an example of what I have in mind. The passage that follows has been made up, but its plaintive sound is so familiar that the reader may find it hard to believe it is not a genuine quotation.

Science is essentially a growth of organised factual knowledge [true or false?], and as science advances, the burden of factual information which it adds to daily is becoming well nigh insupportable. A time will surely come when the scientist must train not for the traditional three or four years, but for ten or more, if he is to equip himself to be a front-line combatant in the battle for knowledge. As things are, the scientist avoids being crushed beneath this factual burden by taking refuge in specialisation, and the increase of specialisation is the distinguishing mark of modern scientific growth. Because of it, scientists are becoming progressively less well able to communicate even with each other, let alone with the outside world; and we must look forward to an ever finer fragmentation of knowledge, in which each specialist will live in a tiny world of his own. St Thomas Aquinas was the last . . .

True or false, all this? False, I should say, in every particular. Science is no more a classified inventory of factual information than history a chronology of dates. The equation of science with *facts* and of the humane arts with *ideas* is one of the shabby genteelisms that bolster up the humanist's self-esteem. That great Platonist, Goldsworthy Lowes Dickinson, who did his best to keep science below stairs, described Aristotle as 'a man of science in the modern sense' because he was 'a careful collector and observer of an enormous range of facts'. No wonder Lowes Dickinson classified Ideas with Philosophy, Art and Love, but the sciences with – *trade*.

The ballast of factual information, so far from being just about to sink us, is growing daily less. The factual burden of a science varies inversely with its degree of maturity. As a science advances, particular facts are comprehended within, and therefore in a sense annihilated by, general statements of steadily increasing explanatory power and compass – whereupon the facts need no longer be known explicitly, that is, spelled out and kept in mind. In all sciences we are being progressively relieved of the burden of singular instances, the tyranny of the particular. We need no longer record the fall of every apple.

Biology before Darwin was almost all facts.[1] My friend R. B. Freeman has brought to light some Victorian examination questions from our oldest English school of zoology, at University College, London. The answers called for nothing more than a voluble pouring forth of factual information.[2] Certainly there is an epoch in the growth of a science during which facts accumulate faster than theories can accommodate them, but biology is over the hump (though biological learned journals still outnumber learned journals of all other kinds by about three to one); and physics is far enough advanced for an eminent

---

[1] Cf. the remarks of S. T. Coleridge, p. 79 below, note 1.

[2] This is one (by no means the longest) of eight questions set by Professor Grant in Comparative Anatomy in February 1860:

'By what special structures are bats enabled to fly through the air? and how do the galeopitheci, the pteromys, the petaurus, and petauristae support themselves in that light element? Compare the structure of the wing of the bat with that of the bird, and with that of the extinct pterodactyl: and explain the structures by which the cobra expands its neck, and the saurian dragon flies through the atmosphere. By what structures do serpents spring from the ground, and fishes and cephalopods leap on deck from the waters? and how do flying-fishes support themselves in the air? Explain the origin, the nature, the mode of construction, and the uses of the fibrous parachutes of arachnidans and larvae, and the cocoons which envelope the young; and describe the skeletal elements which support, and the muscles which move the mesoptera and the metaptera of insects. Describe the structure, the attachments, and the principal varieties of form of the legs of insects; and compare them with the hollow articulated limbs of nereides, and the tubular feet of lumbrici. How are the muscles disposed which move the solid setae of stylaria, the cutaneous investment of ascaris, the tubular peduncle of pentalasmis, the wheels of rotifera, the feet of asterias, the mantle of medusae, and the tubular tentacles of actinae? How do entozoa effect the migrations necessary to their development and metamorphoses? how do the fixed polypifera and porifera distribute their progeny over the ocean? and lastly, how do the microscopic indestructible protozoa spread from lake to lake over the globe?'

physicist to have assured me, with the air of one not wishing to be overheard, that the science itself was drawing to a close . . .

The case for prolonging a scientist's formal education for many years beyond a humanist's follows naturally from the belief that scientific education is a taking on board of specialised technical knowledge. In real life, the time at which a scientist graduates is less important for scientific than for economic and psychological reasons, and for reasons to do with getting enough people through the universities in good time. The length of university schooling is far more important to those whose education ends with graduation than to those for whom education is an indefinitely continued process.

As to scientists' becoming ever narrower and more specialised: the opposite is the case. One of the distinguishing marks of modern science is the disappearance of sectarian loyalties. Newly graduated biologists have wider sympathies today than they had in my day, just as ours were wider than our predecessors'. At the turn of the century an embryologist could still peer down a microscope into a little world of his own. Today he cannot hope to make head or tail of development unless he draws evidence from bacteriology, protozoology, and microbiology generally; he must know the gist of modern theories of protein synthesis and be pretty well up in genetics.

So it is for biologists generally. Isolationism is over; we all depend upon and sustain each other. I must not speak for specialisation in the physical sciences, but feel sure that the continuous and highly successful recruitment of physicists and chemists into biology would not have been possible if they were as specialised as we are often encouraged to believe.

The thoughts I have been criticising are thus not really thoughts at all, but thought-substitutes, declarations of the kind public people make on public occasions when they are desperately hard up for things to say.

Let me turn now to two serious but completely different conceptions of science, embodying two different valuations of scientific life and of the purpose of scientific enquiry. For dialectical reasons I have exaggerated the differences between them, and I do not suggest that anybody cleaves wholly to the one conception or wholly to the other.

According to the first conception, science is above all else an imaginative and exploratory activity, and the scientist is a man taking part in a great intellectual adventure. Intuition is the mainspring of every advancement of learning, and *having ideas* is the scientist's highest accomplishment; the working out of ideas is an important and exacting but yet a lesser occupation. Pure science requires no justification outside itself, and its usefulness has no bearing on its valuation. 'The first man of science', said Coleridge, 'was he who looked into a thing, not to learn whether it could furnish him with food, or shelter, or weapons, or tools, or ornaments, or *play-withs*, but who sought to know it for the gratification of knowing.'

Science and poetry in its widest sense are cognate, as Shelley so rightly said. So conceived, science can flourish only in an atmosphere of complete freedom, protected from the nagging importunities of need and use, because the scientist must travel where his imagination leads him. Even if a man should spend five years getting nowhere, that might represent an honourable and perhaps even a noble endeavour. The patrons of science – today the Research Councils and the great Foundations – should support men, not projects, and individual men rather than teams, for the history of science is for the most part a history of men of genius.

The alternative conception runs something like this: science is above all else a critical and analytical activity; the scientist is pre-eminently a man who requires evidence before he delivers an opinion, and when it comes to evidence he is hard to please. Imagination is a catalyst merely: it can speed thought but cannot start it or give it direction; and imagination must at all times be under the censorship of a dispassionate and sceptical habit of thought. Science and poetry are antithetical, as Shelley so rightly said.[1] Scientific research is intended to enlarge human understanding, and its usefulness is the only objective measure of the degree to which it does so; as to freedom in science, two world wars

[1] This is not a debating point, for Shelley's writing can sustain both views. In his *Defence of Poetry* Shelley defines poetry in a 'universal' sense that comprehends all forms of order and beauty, and includes, therefore, not merely poetry in the narrower sense, but science as well (poetry 'comprehends all science'). Earlier, however, Shelley put Reason and Imagination at opposite poles; if then, as in the second conception I outline, science is regarded as an essentially rational activity, Shelley may quite rightly be allowed to speak for the view that science and poetry are antithetical. See D. G. King-Hele in the *New Scientist* **14**, 352–4, 1962; and Graham Wallas, *The Art of Thought* (London, 1926).

have shown us how very well science can flourish under the pressures of necessity. Patrons of science who really know their business will support projects, not people, and most of these projects will be carried out by teams rather than by individuals, because modern science calls for a consortium of the talents and the day of the individual is almost done. If any scientist should spend five years getting nowhere, his ambitions should be turned in some other direction without delay.

I have made the one conception a little more romantic than it really is, and the other a little more worldly, and to restore the balance, I want to express the distinction in a different and, I think, more fundamental way.

In the romantic conception, truth takes shape in the mind of the observer: it is his imaginative grasp of *what might be true* that provides the incentive for finding out, so far as he can, what *is* true. Every advance in science is therefore the outcome of a speculative adventure, an excursion into the unknown. According to the opposite view, truth resides in nature and is to be got at only through the evidence of the senses: apprehension leads by a direct pathway to comprehension, and the scientist's task is essentially one of *discernment*. This act of discernment can be carried out according to a Method which, though imagination can help it, does not depend on the imagination: the Scientific Method will see him through.[1]

Inasmuch as these two sets of opinions contradict each other flatly in every particular, it seems hardly possible that they should both be true; but anyone who has actually done or reflected deeply upon scientific research knows that there is in fact a great deal of truth in both of them. For a scientist must indeed be freely imaginative and yet sceptical, creative and yet a critic. There is a sense in which he must be free, but another in which his thought must be very precisely regimented; there is poetry in science, but also a lot of bookkeeping.

[1] For the conception that *truth is manifest*, see the critical analysis by Karl Popper, 'On the Sources of Knowledge and of Ignorance', in *Conjectures and Refutations* (London, 1963). The question where Truth resides can also be put of Beauty, and answered in the same two ways, for the romantic view does not distinguish them. For the history of the idea that *beauty is manifest* (as opposed to being in the eye of the observer), see Logan Pearsall Smith, *The Romantic History of Four Words: romantic, originality, creative, genius*, S.P.E. Tract 17 (Oxford, 1924).

There is no paradox here: it just so happens that what are usually thought of as two alternative and indeed competing accounts of *one* process of thought are in fact accounts of the *two* successive and complementary episodes of thought that occur in every advance of scientific understanding. Unfortunately, we in England have been brought up to believe that scientific discovery turns upon the use of a method analogous to and of the same logical stature as deduction, namely the method of *Induction* – a logically mechanised process of thought which, starting from simple declarations of fact arising out of the evidence of the senses, can lead us with certainty to the truth of general laws. This would be an intellectually disabling belief if anyone actually believed it, and it is one for which John Stuart Mill's methodology of science must take most of the blame. The chief weakness of Millian induction was its failure to distinguish between the acts of mind involved in discovery and in proof. It was an understandable mistake, because in the process of deduction, the paradigm of all exact and conclusive reasoning, discovery and proof may depend on the same act of mind: starting from true premises, we can derive and so 'discover' a theorem by reasoning which (if it has been carried out according to the rules) itself shows that the theorem must be true. Mill thought that his process of 'induction' could fulfil the same two functions; but, alas, mistakenly, for it is not the origin but only the *acceptance* of hypotheses that depends upon the authority of logic.

If we abandon the idea of induction and draw a clear distinction between *having an idea* and *testing it* or *trying it out* – it is as simple as that, though it can be put more grandly – then the antitheses I have been discussing fade away. Obviously 'having an idea' or framing a hypothesis is an imaginative exploit of some kind, the work of a single mind; obviously 'trying it out' must be a ruthlessly critical process to which many skills and many hands may contribute. The form taken by scientific criticism is obvious too: experimentation *is* criticism; that is, experimentation in the modern sense, according to which an experiment is an act performed to test a hypothesis, not in the old Baconian sense, in which an experiment was a contrived experience intended to enlarge our knowledge of what actually went on in nature. Bacon exhorted us, rightly too, not to speculate upon but actually to experiment with loadstone and burning glass and rubbed amber; *his* experiments answer the question 'I wonder what would happen if . . .?' Baconian experimentation is not a critical activity but a kind of creative play.

The distinction between – and the formal separateness of – the creative and the critical components of scientific thinking is shown up by logical dissection, but it is far from obvious in practice because the two work in a rapid reciprocation of guesswork and checkwork, proposal and disposal, *Conjecture and Refutation*. Though imaginative thought and criticism are equally necessary to a scientist, they are often very unequally developed in any one man. Professional judgement frowns upon extremes. The scientist who devotes his time to showing up the inadequacies of the work of others is suspected of lacking ideas of his own, and everyone soon loses patience with the man who bubbles over with ideas which he loses interest in and fails to follow up.

The general conception of science which reconciles, indeed literally joins together, the two sets of contradictory opinions I have just outlined is sometimes called the 'hypothetico-deductive' conception. For our present clear understanding of the logical structure and wider scientific implications of the hypothetico-deductive system we are of course indebted to Karl Popper's *Logik der Forschung* of 1934, translated into English as *The Logic of Scientific Discovery*.[1]

Everything I have said so far about the hypothetico-deductive system applies with exactly the same force to 'applied' science, even in its simplest and most familiar forms, as to that which is commonly called 'pure' or 'basic'. Imaginative conjecture and criticism, in that order, underlie the physician's diagnosis of his patient's ailments or the mechanic's explanation of why a car won't run. The physician may like to think himself, as Darwin did, an inductivist and a good Baconian, but with equally little reason, for Darwin was no inductivist; no more is he.

What now follows is an attempt to analyse, not the difference between basic and applied science, but the motives which have led people to think it highly important, and above all to make it the basis of an intellectual class-distinction.

Francis Bacon was not the first to distinguish basic from applied

---

[1] For earlier accounts of the hypothetico-deductive scheme, see 'Hypothesis and Imagination' below, pp. 115–35, especially pp. 125–8.

science, but no one before him put the matter so clearly and insistently, and the distinction as he draws it is unquestionably just. 'It is an error of special note,' said Bacon, pondering upon the many infirmities of current learning, 'that the industry bestowed upon experiments hath presently, upon the first access into the business, seized upon some designed operation; I mean sought after *Experiments of Use* and not *Experiments of Light and Discovery*.'[1] (The image of light is a favourite of Bacon's, and the idea of kindling a light in nature.) Bacon's distinction is between research that increases our power over nature and research that increases our understanding of nature, and he is telling us that the power comes from the understanding. He felt his distinction upheld by the example of that Divine Builder who created Light only, and no *Materiate Work*, in the first day, turning to what we should nowadays presumably call Applied Science in the days following.

No one now questions Bacon's argument. Who nowadays would try to build an aeroplane without trying to master the appropriate aerodynamic theory? Sciences not yet underpinned by theory are not much more than kitchen arts. Aeronautics, and the engineering and applied sciences generally, do of course obey the Baconian ruling that what is done for use should so far as possible be done in the light of understanding. Unhappily, Bacon's distinction is not the one we now make when we differentiate between the basic and applied sciences. The notion of *purity* has somehow been superimposed upon it, and in a new usage that connotes a conscious and inexplicably self-righteous disengagement from the pressures of necessity and use. The distinction is not now between the empirically founded sciences and those whose axioms were supposedly known a priori; rather it is between polite and rude learning, between the laudably useless and the vulgarly applied, the free and the intellectually compromised, the poetic and the mundane.

Let me first say that all this is terribly, terribly English, or anyhow Anglo-Saxon. Making pure and applied science the basis of a class distinction helps us to forget that it was our engineers and merchants, not the armed forces, the Civil Service and the gentry, who won for us that very grand position in the world from which we have now stepped down. It is not always easy to explain to foreigners the whole

[1] *The Great Instauration*, Preface.

connotation of 'pure' in the context 'pure research'. They only shake their heads uneasily and wonder if it may not have something to do with cricket. They lack also our Own Very Special Blend of high-mindedness and humbug in that reasoning which champions Pure Research because, while it enables the human spirit to breathe freely in the thin and serene atmosphere of the intellectual highlands, it is also a splendid long-term investment. Invest in applied science for quick returns (the spiritual message runs), but in pure science for capital appreciation.[1] And so we make a special virtue of encouraging pure research in, say, cancer institutes or institutes devoted to the study of rheumatism or the allergies – always in the hope, of course, that the various lines of research, like the lines of perspective, will converge somewhere upon a point. But there is nothing virtuous about it! We encourage pure research in these situations because we know no other way to go about it. If we knew of a direct pathway leading to the solution of the clinical problem of rheumatoid arthritis, can anyone seriously believe that we should not take it?

The more creditable part of our English reverence for pure research derives, I believe, from a certain accident of our aesthetic history. Let us concede that imaginative thought plays an important part – no matter what – in discovery and invention. Now in this country the quintessential form of imaginative activity has always been poetic invention. Hereabouts, a man inspired is typically a poet inspired. Unfortunately there is no such thing as Applied Poetry – or rather, there is, but we think little of it. We look askance at poetry for the occasion, even for Royal occasions. For poetry 'is not like reasoning, a power to be exerted according to the determination of the will. A man cannot say "I will compose poetry." The greatest poet even cannot say it.' Still less can he say that he will compose joyful or lugubrious poetry, or poetry upon a given theme. 'Poetry . . . is not subject to the control of the active powers of the mind, and its birth and recurrence have no necessary connection with the consciousness or will.'

Substitute 'pure science' for 'poetry' in Shelley's manifesto, and it

[1] V. B. Wigglesworth has pointed out that in the first edition of his *Grammar of Science* (1892) Karl Pearson chose Hertzian waves, essentially radio waves, as an example of a discovery of no apparent usefulness. This is the best example I know of the apparently useless bringing in the goods (as Pearson thought it probably would).

will help us to understand the aesthetic conspiracy which has led us to think so much more highly of pure research than of research with an acknowledged practical purpose. It was quite otherwise in the early days of the Royal Society, when there was a danger that any experiment not immediately useful would be dismissed as play.

> It is strange [said Thomas Sprat][1] that we are not able to inculcate into the minds of many men, the necessity of that distinction of my Lord Bacon's, that there ought to be Experiments of Light, as well as of Fruit . . . If they will per-sist in contemning all Experiments, except those that bring with them immedi-ate gain, and a present Harvest, they may as well cavil at the Providence of God, that he has not made all the seasons of the year to be times of mowing, reaping, and vintage.

Sprat is not arguing for pure research in the sense in which we should now use that term, but rather against a hasty opportunism; his formula is 'Light now, for Use hereafter'.

In countries in which poetry is not the top art form, the idea of occasional or commissioned art is commonplace and honourable, and there is correspondingly less fuss, if any, about the distinction between pure science and applied. Tapestry and statuary, stained glass, murals and portraiture, palaces, cathedrals and town halls, fire music, water music and funeral marches – most are commissioned, and the act of being commissioned may itself light up the imagination.[2] For everyone who uses imagination knows that it can be trained and guided and deliberately stocked with things to be imaginative about. Only the irremediably romantic can believe, as Coleridge did, that artistic creation is a microcosmic version of that Divine sort of creation which can make something out of nothing, or out of a homogeneous cloud of forms or notions – and how little right *he* had to think so has been made clear by *The Road to Xanadu* and other aetiological studies of Coleridge's choice of images and words. To the sober-minded the 'spontaneity' of an idea signifies nothing more than our unawareness of what preceded its irruption into conscious thought.

I am labouring these obvious points in order to make it clear that

---

[1] Thomas Sprat, *History of the Royal Society* (1667), p. 245. See also the Intro-duction to this volume, p. 14.

[2] Benjamin Britten always spoke in favour of occasional music: see his speech *On Receiving the first Aspen Award* (London, 1964).

poetic inspiration is not a valid guide to imaginative activity in all its forms; that there is no case for looking down on commissioned art or science or on extra-mural sources of inspiration; and that our English reverence for Pure Research, though historically understandable, and perhaps even lovable, is also slightly ridiculous.

If we study the criteria that underlie a scientist's own valuation of science, we shall certainly not find purity among them. This is a significant omission, for the scientific valuation of scientific research is remarkably uniform throughout the world.

Here then are some of the criteria used by scientists when judging their colleagues' discoveries and the interpretations put upon them. Foremost is their *explanatory value* – their rank in the grand hierarchy of explanations and their power to establish new pedigrees of research and reasoning. A second is their clarifying power, the degree to which they resolve what has hitherto been perplexing; a third, the feat of originality involved in the research, the surprisingness of the solution to which it led, and so on. Scientists give weight (though much less weight than mathematicians do) to the elegance of a solution and the economy of the thought and work that went into it; they give credit, too, for the difficulty of the enterprise as a whole – the size of the obstacles that had to be got over or got round before the solution was reached. But purity, as such, is nowhere. Nor is usefulness, which has its own scale of valuation and its own rewards. Let usage guide us: 'How neat!' one scientist might say of another's work – or 'How ingenious!' – or 'How very illuminating!' – but never, in my hearing anyway, 'How pure!'

There is without doubt a case for uncommitted or disengaged research, but it is not self-evident, and there may turn out to be better ways of doing what pure research professes to do. For example, in institutes of basic research it is believed and hoped that something practically useful may be come upon in the course of free-ranging enquiry, whereupon research which has hitherto shed diffuse light will now come sharply into focus. This procedure works; that is, it works sometimes, and it may be the best we can do, but there's no knowing, for alternative approaches have not yet been tried out on a sufficiently large scale. Might not the converse approach be equally effective, given equal opportunity and equal talent? – to start with a concrete problem, but then to allow the research to open out in the direction of greater generality, so that the more particular and special discoveries can be made to rank as theorems derived from statements of higher explanatory value. I can see no reason why this approach, *if it were to be attempted by persons of the same ability*, should not work just as well as its more conventional alternative; in fact I believe that some great American companies are moving towards it and already have some brilliant achievements to justify their choice – for example, the growth of a generalised communications theory out of the practical problems of sending messages by telephone. Research done in this style is always in focus, and those who carry it out, if temporarily baffled, can always retreat from the general into the particular.

If our reverence for Pure Science is a rather parochial thing, a by-product of the literary propaganda of the romantic revival; if no case can be made for it on philosophic grounds; if purity is not part of a scientist's own valuation of science; then why on earth do we think so highly of it? It is, I think, our humanist brethren who have taught us to believe that, while pure science is a genteel and even creditable activity for scientists in universities, applied science, with all its horrid connotations of trade, has no place on the campus; for only the purest of pure science can give countenance to research in the humanities – research which, though it cannot very well be described as pure, for want of anything applied to compare it with, can all too readily be described as useless. The humanist fears that if we abandon the ideal of pure knowledge, knowledge acquired for its own sake, then usefulness becomes the only measure of merit; and that if it does become so, research in the humane arts is doomed.

These fears, I have tried to explain, are groundless. Neither its purity *nor* its usefulness enter a scientist's valuation of his own research. The scientist values research by the size of its contribution to that huge, logically articulated structure of ideas which is already, though not yet half built, the most glorious accomplishment of mankind. The humanist must value his research by different but equally honourable standards, particularly by the contribution it makes, directly or indirectly, to our understanding of human nature and conduct, and human sensibility.

I have been trying to make a case for a critical study of the organisation of research, by which I do *not* mean either the allocation of administrative responsibilities for research or the economics and logistics of science,[1] important though they are. I mean a study of the behavioural and intellectual structure of everything that goes into the enlargement of our knowledge and understanding of nature. I have already mentioned a few of the problems that might be on the agenda of such an investigation, and here are a few more. Are scientists a homogeneous body of people in respect of temperament, motivation, and style of thought? (Obviously not: but we talk of *the* scientist nevertheless.) Is there such a thing a 'scientific mind'? I think not. Or *the* scientific method? Again, I think not. What exactly are the terms of a scientist's contract with the truth? This is an important question, for according to the interpretation of the scientific process which I myself think the most plausible, a scientist, so far from being a man who never knowingly departs from the truth, is always *telling stories* in a sense not so very far removed from that of the nursery euphemism – stories which might be about real life but which must be tested very scrupulously to find out if indeed they are so.[2]

Again, and in no particular order: Is it really true that a good or genuine scientist is, or should be, indifferent to matters of priority, caring only for the Advancement of Learning and nothing for who causes it to come about? How can the *frettoso* of research be combined harmoniously with the *adagio* of administration? How can the productivity of scientists be increased: is full-time research really a good thing for more than a lucky or slightly obsessional minority, and, if not, what else should a scientist do, and how should his time be parcelled out to best advantage?

These are important questions, and their answers must no longer be entrusted to asseveration – to 'peremptory fits of asseveration', Bacon said, when clearing the ground for his own *Great Instauration*. They

---

[1] See M. Goldsmith and A. L. Mackay (eds), *The Science of Science* (London, 1964).

[2] A 'story' is more than a hypothesis: it is a theory, a hypothesis together with what follows from it and goes with it, and it has the clear connotation of completeness within its own limits. I notice that laboratory jargon follows this usage, e.g. 'Let's get So-and-so to tell his story about' something or other, an invitation which So-and-so may decline on the grounds that his work 'doesn't make a story yet' or accept because he 'thinks he's got a story'. There is a slightly depreciatory flavour about this use of 'story' because fancy has to be used to fill in the gaps and some people tend to overdo it.

will have to be thought over and argued out with some sense of urgency; and we here in England had better be quick about it, in case the wind changes and we get fixed permanently in our Anglo-Saxon attitudes to research.

SCIENCE AND LITERATURE

I

I HOPE I shall not be thought ungracious if I say at the outset that
nothing on earth would have induced me to attend the kind of lecture
you may think I am about to give. Science and literature – what a
hackneyed subject, you must feel. Must we go into that again? What
can there be to say that has not already been very well said by I. A.
Richards, Aldous Huxley, C. P. Snow, Martin Green, J. Bronowski,
D. G. King-Hele and half a dozen others?[1]

Let me begin with an outline of some of the things I do *not* intend to
say. I shall say nothing whatsoever about education, and have no
formula for compounding science and literature into a single diet; nor
shall I say, or even be thinking, that imaginative literature can be regarded as an antidote or counter-irritant to science, or vice versa.
There will be no readings from poetry written by scientists – not even
a quotation or autopsy specimen from the poetry of George John
Romanes, FRS.[2] I shall not declare that henceforward the discoveries,
ideas and adventures of science should become a bigger part of the
subject-matter of poetry, as Wordsworth[3] thought they might; nor
shall I reproach 'poets and magazine critics', as Peacock did,[4] with
carrying on just as if 'there were no such things in existence as mathematicians, astronomers, chemists, moralists, metaphysicians, historians,
politicians, and political economists'.

[1] I shall quote below from I. A. Richards, *Science and Poetry* (London, 1926);
Aldous Huxley, *Literature and Science* (London, 1963). See also C. P. Snow, *The
Two Cultures: And a Second Look* (Cambridge, 1964); Martin Green, *Science and
the Shabby Curate of Poetry* (London, 1964); J. Bronowski, *Science and Human
Values*, revised ed. (New York, 1965); D. G. King-Hele, *Shelley: His Thought and
Work* (London, 1960); *Erasmus Darwin* (London, 1963).
[2] See *A Selection from the Poems of George John Romanes*, ed. T. Herbert Warren
(London, 1896). I am much obliged to Mr R. B. Freeman for calling my attention
to Romanes's poetry. I have never read worse.
[3] See Wordsworth's introduction to the *Lyrical Ballads* (1802 edition).
[4] *The Four Ages of Poetry* (1820).

After these various abjurations, what is there left to say? If I had to choose a motto for this lecture, I should turn a remark of Lowes Dickinson's upside down. 'When science arrives,' said Lowes Dickinson,[1] 'it expels literature' – an echo, perhaps, of Keats's lament that science unweaves the rainbow and makes a dull ordinariness out of awful things. The case I shall find evidence for is that when literature arrives, it expels science. There are large territories of human belief and learning upon which both science and literature have very important things to say, for example, social and cultural anthropology, psychology and human behaviour generally, and even cosmology. These subjects lie within the compass of literature in so far as they have to do with human hopes, fears, beliefs and motives; with the attempt to give an account of ourselves and investigate our condition; and with matters of general culture, by which I mean the whole pattern of the way in which people think and carry on. The case that can be made for science is that in all these subjects we have *also* to work towards a special kind of understanding which, though imaginative in origin (as I shall hope to convince you), is under the censorship or restrictive influence of a certain kind of obligation towards the truth.

The way things are at present, it is simply no good pretending that science and literature represent complementary and mutually sustaining endeavours to reach a common goal. On the contrary, where they might be expected to co-operate, they compete. I regret this very much, don't think it necessary, and wish it were otherwise. We are going through a bad episode in cultural history. We all want to be friends, and one day perhaps we shall be so. 'Let us advance together, men of letters and men of science,' said Aldous Huxley,[2] 'further and further into the ever expanding regions of the unknown.' That is a fine ambition, though most of us will feel awkward at its wording; but if it is to be achieved, scientists and men of letters must work their way towards an understanding – not just of each other's accomplishments (there is mischief and magnificence in both), nor just of each other's purposes (which are doubtless mixed, though officially both are good), but of each other's methods and energising concepts and the quality and pattern of movement of each other's thought. I want therefore to

[1] *Plato and His Dialogues* (London, 1931).
[2] op. cit. (p. 42 above, note 1).

discuss imagination and criticism as they enter into science on the one hand and into literature on the other; to explain why I think that scientific and literary conceptions of style and matters of communication cannot be reconciled; and finally to compare scientific and poetic notions of the truth. Towards the end, I shall use Freudian psychoanalysis and existential psychiatry to illustrate the way in which science and literature compete for the territories on which they both have claims.

2

Let me begin by discussing the character and interaction of imagination and critical reasoning in literature and in science. I shall use 'imagination' in a modern sense (modern on the literary time scale, I mean), or, at all events, in a sense fully differentiated from mere fancy or whimsical inventiveness. (It is worth remembering that when the phrase 'creative imagination' is used today, we are expected to look solemn and attentive, but in the eighteenth century we could as readily have looked contemptuous or even shocked.)

The official Romantic view is that Reason and the Imagination are antithetical, or at best that they provide alternative pathways leading to the truth, the pathway of Reason being long and winding and stopping short of the summit, so that while Reason is breathing heavily there is Imagination capering lightly up the hill. It is true that Shelley[1] recognised a poetical element in science, though 'the poetry is concealed by the accumulation of facts and calculating processes'; true also that in one passage of his famous rhapsody, he was kind enough to say that poetry comprehends all science – though here, as he makes plain, he is using poetry in a general sense to stand for all exercises of the creative spirit, a sense that comprehends imaginative literature itself as one of its special instances. But in the ordinary usages to which I shall restrict myself, Reason and Imagination are antithetical. That was Shelley's view and Keats's, Wordsworth's and Coleridge's; it was also Peacock's, for whom Reason was marching into territories formerly occupied by poets; and it was also the view of William Blake,[2] who

[1] Percy Bysshe Shelley, *A Defence of Poetry* (1821).

[2] William Blake, *Milton* (1804), book 2, plate 41; and *Jerusalem* (1804), chapter 1, plate 10.

came 'in the grandeur of Inspiration to cast off Rational Demonstration . . . to cast off Bacon, Locke, & Newton'; 'I will not Reason &
Compare – my business is to create.'

This was not only the official view of the Romantic poets; it was also
the official scientific view. When Newton wrote *Hypotheses non fingo*,
he was taken to mean that he reprobated the exercise of the imagination in science. (He did not 'really' mean this, of course, but the importance of his disclaimer lies precisely in this misunderstanding of it.)
Bacon too, and later on John Stuart Mill, were taken as official spokesmen for the belief that there existed, or could be devised, a calculus of
discovery, a formulary of intellectual behaviour, which could be relied
upon to conduct the scientist towards the truth, and this new calculus
was thought of almost as an antidote to the imagination, as it had been
in Bacon's own day an antidote to what Macaulay[1] called the 'sterile
exuberance' of scholastic thought. Even today this central canon of
inductivism – that scientific thought is fully accountable to reason – is
assumed quite unthinkingly to be true. 'Science is a matter of disinterested observation, unprejudiced insight and experimentation, and
patient ratiocination within some system of logically correlated concepts' – an important opinion, for Aldous Huxley[2] is a man thought to
speak with equal authority for science and letters.

Huxley would, of course, have been the last man to deny imagination
a role in science; nor even did the high priest of Inductivism, Karl
Pearson; but in science a creative imagination is the privilege of the
rare spirit who achieves in a blaze of intuition what the rest of us can
only do by rote or by 'analytic industry' (Wordsworth's term). But
the point is (I am still recounting the official view) that we *can* do it;
we may not all be great cooks, but we can all read the instructions on
the packet. There *is* a calculus of discovery, and it works independently
of intuition, though nothing like so fast.

The reductionist view – of the *complete* accountability of science to
reason – is no longer believed in by most people who have thought
deeply about the nature of the scientific process. An entirely different
conception grew up in the writings of William Whewell, Stanley

[1] Thomas Babington Macaulay, *Lord Bacon* (1837), an extended review of
Montagu's edition of Bacon's works that first appeared in the *Edinburgh Review*.
[2] op. cit. (p. 42 above, note 1).

Jevons, C. S. Peirce, and latterly of Karl Popper.[1] Because its message is in danger of being lost in technical discussion over points of detail, let me explain the gist of it in very general terms.

All advances of scientific understanding, at every level, begin with a speculative adventure, an imaginative preconception *of what might be true* – a preconception that always, and necessarily, goes a little way (sometimes a long way) beyond anything which we have logical or factual authority to believe in. It is the invention of a possible world, or of a tiny fraction of that world. The conjecture is then exposed to criticism to find out whether or not that imagined world is anything like the real one. Scientific reasoning is therefore at all levels an inter-action between two episodes of thought – a dialogue between two voices, the one imaginative and the other critical; a dialogue, as I have put it, between the possible and the actual, between proposal and dis-posal, conjecture and criticism, between what might be true and what is in fact the case.

In this conception of the scientific process, imagination and criticism are integrally combined. Imagination without criticism may burst out into a comic profusion of grandiose and silly notions. Critical reason-ing, considered alone, is barren. The Romantics believed that poetry, *poiesis*, the creative exploit, was the very opposite of analytic reasoning, something lying far above the common transactions of reason with reality. And so they missed one of the very greatest of all discoveries, of the synergism between imagination and reasoning, between the inventive and the critical faculties.[2] I call it a 'discovery', but no one person made it. Coleridge could have made it; he alone in a hundred and fifty years was qualified in every way to do so. It is a tragedy of cultural history that he did not.

At this point a spokesman for literature might say, 'I accept the idea that scientific reasoning can be resolved into a dialogue between

---

[1] The appropriate references are given elsewhere in this volume, principally in 'Induction and Intuition in Scientific Thought' and 'Hypothesis and Imagination'.

[2] After I delivered this lecture, Sir Ewart Jones called my attention to J. H. Van't Hoff's highly original inaugural lecture on 'Imagination in Science', given in 1878 when he was twenty-six years old. Van't Hoff speaks of the 'synergism of imagination and critical judgement', though not quite in the sense I intended here. His lecture has been translated and annotated by G. F. Springer (Berlin, 1967).

critical and inventive faculties, or something of that general nature, but what is so distinctively scientific about it, and why should it be held to distinguish science from imaginative literature?' He would quote Matthew Arnold,[1] perhaps – 'all poetry is criticism of life' – but in this context he would probably not wish to press the point, for Arnold, too, saw criticism and inventiveness as antithetical, and when he says, for example, that 'the critical power is of lower rank than the creative', he shows that he has no idea of the existence of forms of intellection to which class distinctions of this kind do not apply. But would it not be reasonable to say that literary criticism has a function cognate with that which I have attributed to criticism in science.?

I am not deeply enough read in modern critical literature to say whether there is anything in such an argument or not, but my inclination would be to say that there is not. I have been saying that the critical episode of thought is something integral with scientific reasoning; that which has not yet been exposed to it is not yet science. Literary criticism, on the other hand, is a branch of literature which has literature as its subject-matter, and that is an altogether different thing. A better case could be made for saying that literary criticism has something in common with scientific methodology. There is something in this, surely, but the similarities between them are rather dull and the differences interesting. Scientific methodology has to do with matters of validation and justification as they enter into all forms of scientific thinking, with the trustworthiness of evidence, and with the analysis of certain formal ideas that are common to all the sciences, for example, the ideas of causality or of reducibility and emergence. But it has nothing to do with the motives and purposes of scientists or with the degree to which their work achieves them: science is known to us in terms of accomplishment, not in terms of endeavour. It does not attempt to justify science in any sense except the scientific; above all it does not try to see scientific thought and action as elements of general culture. What *should* be the equivalent in science of literary criticism is therefore represented by a great emptiness which is a reproach to all scholars, scientists and humanists alike. I cannot even think of a name for the new discipline that might fill those empty spaces, for 'scientific

---

[1] *The Study of Poetry* and *The Function of Criticism at the Present Time*, reprinted in *Essays in Criticism*, first series (1865) and second series (1888).

humanism' is too deeply committed to a different meaning, and its practitioner, the 'scientific humanist', has too long and too often been made a figure of fun. (You have only to think of some of Peacock's caricatures, or of Sir Austin Feverel.)[1] Perhaps the word I want is 'criticism' itself, without qualification.

The gist of what I have been saying is this. Our traditional views about imagination and criticism in literature and in science are based upon the literary propaganda of the Romantic poets and the erroneous opinions of inductivist philosophers. Imagination is the energising force of science as well as of poetry, but in science imagination and a critical evaluation of its products are integrally combined. To adopt a conciliatory attitude, let us say that science is that form of poetry (going back now to its classical and more general sense) in which reason and imagination act together synergistically. This simple formal property (which can, of course, be set out in a much more professional and specific language than anything I have attempted here) represents the most important methodological discovery of modern thought.

3

I now turn to a discussion of matters of style.

The poet 'yieldeth to the powers of the mind an image of that whereof the philosopher bestoweth but a wordish description, which doth neither strike, pierce, nor possess the sight of the soul'.[2] At a time when the writing of English had reached a peak of adventurousness and effulgence – and partly, but not wholly, because it had done so – the New Philosophers of the seventeenth century (new scientists, we should now say) were put officially on their guard against the danger of being carried away by the sound of their own voices. I say 'officially' because the warning came from the Royal Society. Then and for evermore they were to abjure the 'painted scenes and pageants of the brain'.[3] Their writing was 'manly and yet plain . . . It is not broken by

---

[1] In George Meredith's *The Ordeal of Richard Feverel* (1859), the tiresome Sir Austin is explicitly described as a 'scientific humanist', the first example of this particular usage I have come across.

[2] Sir Philip Sidney, *An Apology for Poetry* (1595).

[3] Abraham Cowley, *To the Royal Society* (1663).

ends of Latin, nor impertinent quotations . . . not rendered intricate by long parentheses, nor gaudy by flaunting metaphors; not tedious by wide fetches and circumferences of speech'. The scientific style was to be 'as polite and as fast as marble'. I am quoting Joseph Glanvill, FRS[1] – not a good one to talk, perhaps, as his own style was described by H. Oldenburg as somewhat florid; but he spoke for common opinion. Not Words but Works were to carry the message of the New Philosophy. 'We believe a scientist because he can substantiate his remarks,' said I. A. Richards, 'not because he is eloquent and forcible in his enunciation. In fact, we distrust him when he seems to be influencing us by his manner.'[2] There is a passage in a still undiscovered manuscript of Bacon's in which he expresses his abhorrence of the *venditio suavis*, or soft sell.

Lowes Dickinson was right, if we take him in a narrower sense than he may have intended. Science and imaginative *writing* are utterly incongruous, in English anyway – the French tradition is more permissive – and the effect of combining them is merely absurd. In science the imaginative element lies in the conception, and not at all in the language by which the conception is made explicit or is conveyed. (The 'language' might indeed use the symbolism of chemistry, mathematics or electronic circuitry.) Clarity can be, *must* be achieved, and with a natural stylist like D'Arcy Thompson, grace. But a scientist's fingers, unlike a historian's, must never stray toward the diapason, and a falling cadence is allowed only to mark, and perhaps be the welcome evidence of, the end of a 'presentation'.

By the time of the New Philosophy, the competition or disputation between eloquence and wisdom, style and substance, medium and message, had already been in progress for nearly two thousand years, but as far as the New Philosophy was concerned, the Royal Society, with the formidable support of John Locke and Thomas Hobbes, may be thought to have settled the matter once and for all: scientific and philosophic writing were on no account to be made the subject of a literary spectacle and of exercises in the high rhetoric style.

[1] *Plus Ultra* (1668). I learned Henry Oldenburg's opinion from Jackson Cope's introduction to a modern facsimile edition (Gainesville, 1958).
[2] op. cit. (p. 42 above, note 1).

This position has been threatened only during those two periods in which our native philosophic style (which is also a style of thinking) was obfuscated by influences from abroad. During the Gothic period of philosophic writing, which began before the middle of the nineteenth century and continued until the First World War, we were all oppressed and perhaps mildly stupefied by metaphysical profundities of German origin. But although those tuba notes from the depths of the Rhine filled us with thoughts of great solemnity and confusion, it was not as music, thank heavens, that we were expected to admire them. The style was not an object of admiration in itself. Today, though we are now much better armed against it, speculative metaphysics has given way to what might be called a *salon* philosophy as the chief exotic influence, and French writers enjoy the reverential attention that was at one time thought due to German. Style has now become an object of first importance, and what a style it is! For me it has a prancing, high-stepping quality, full of self-importance; elevated indeed, but in the balletic manner, and stopping from time to time in studied attitudes, as if awaiting an outburst of applause. It has had a deplorable influence on the quality of modern thought in philosophy and in the behavioural and 'human' sciences.

The style I am speaking of, like the one it superseded, is often marked by its lack of clarity, and for this reason we are apt to complain that it is sometimes very hard to follow. To say as much, however, may now be taken as a sign of eroded sensibilities. I could quote evidence of the beginnings of a whispering campaign against the virtues of clarity. A writer on structuralism in the *Times Literary Supplement* has suggested that thoughts which are confused and tortuous by reason of their profundity are most appropriately expressed in prose that is deliberately unclear. What a preposterously silly idea! I am reminded of an airraid warden in wartime Oxford who, when bright moonlight seemed to be defeating the spirit of the blackout, exhorted us to wear dark glasses. He, however, was being funny on purpose.

I must not speak of obscurity as if it existed in just one species. A man may indeed write obscurely when he is struggling to resolve problems of great intrinsic difficulty. This was the obscurity of Kant, one of the greatest of all thinkers. There is no more moving or touching passage in his writings than that in which he confesses that he has no gift for lucid exposition, and expresses the hope that in due course others will help to make his intentions plain.[1]

In the eighteenth century obscurity was regarded as a disfigurement not merely of philosophic and scientific but also of theological prose. To conceal meaning (it was reasoned) is equally to conceal lack of meaning, so we don't know where we stand. George Campbell[2] (the Scottish philosopher and divine, not the poet) thought himself specially afflicted by mystical theology, and his interpretation of it will pass very well today. Mystical theology is a prose-offering to the Almighty; and just as it is in the nature of a living sacrifice that it should be deprived of life before being offered up to the Godhead, so a prose-offering must be deprived of – sense. This then is constitutive obscurity: that which appears to be nonsense for the simplest of all reasons, namely, that it is not sense.

But even in those enlightened days, the appeal of obscurity was clearly recognised. Of Dryden – even of Dryden – Johnson said that 'he delighted to tread upon the brink of meaning, where light and darkness begin to mingle'.[3] Don't we all, up to a point? We all recognise a voluptuary element in the higher forms of incomprehension and a sense of deprivation when matters which have hitherto been mysterious are now made clear.

The *rhetorical* use of obscurity is, however, a vice. It is often said – and it was said of Kant[4] – that the purpose of obscure or difficult writing is to create the illusion of profundity, and the accusation need not be thought an unjust one merely because it is trite. But in its more subtle usages, obscurity can be used to create the illusion of a deeply reasoned discourse. Suppose we read a text with a closely reasoned argument which is complex and hard to follow. We struggle with it, and as we go along we may say, 'I don't see how he makes that out', or 'I can see now what he's getting at', and in the end we shall probably get there, and either agree with what the author says or find reasons for taking a different view. But suppose there is no argument; suppose that the text is asseverative in manner, perhaps because analytical

[1] Immanuel Kant, *Critique of Pure Reason*, introduction to the second edition (1787).
[2] *The Philosophy of Rhetoric* (1776).
[3] 'Dryden', in *The Lives of the Poets*, vol. 2 (1781).
[4] See p. 22 above, note 1.

reasoning has been repudiated in favour of reasoning of some higher kind. If now the text is made hard to follow because of *non sequiturs*, digressions, paradoxes, impressive-sounding references to Gödel, Wittgenstein and topology, 'in' jokes, and a general determination to keep all vulgar sensibilities at bay, then again we shall have great difficulty in finding out what the author intends us to understand. We shall have to reason it out therefore, much as we reasoned out Latin unseens or a passage in some language we didn't fully understand. In both texts some pretty strenuous reasoning may be interposed between the author's conceptions and our understanding of them, and it is strangely easy to forget that in one case the reasoning was the author's but in the other case our own. We have thus been the victims of a confidence trick.

Let me end this section with a declaration of my own. In all territories of thought which science or philosophy can lay claim to, including those upon which literature has also a proper claim, no one who has something original or important to say will willingly run the risk of being misunderstood; people who write obscurely are either unskilled in writing or up to mischief. The writers I am speaking of are, however, in a purely literary sense, extremely skilled.

4

Let me now turn to a comparison between scientific and poetic notions of the truth, though only as far as it may help to recognise and define the literary syndrome in scientific or quasi-scientific thought.

When the word is used in a scientific context, *truth* means, of course, correspondence with reality. Something is true which is 'actually' true, is indeed the case. This is empirical truth – truth in the sense in which it is true to say that I am at this moment delivering the Romanes Lecture and not standing on my head on an ice floe in the North Atlantic; and you know that correspondence with reality in just this sense is the test that all scientific theories must be put to, no matter how lofty or how trivial they may be.

We must at once dismiss the idea that empirical or factual truth as scientists use it (or lawyers or historians) is an elementary or primitive notion of which everyone must have an intuitive or inborn understanding. On the contrary, it is very advanced, very grown-up, something we learn to appreciate, not something that comes to us naturally. We must also, I think, dismiss the inductive interpretation of the way in which truth enters into scientific enquiry. In classical inductive theories of scientific method, plain factual truth is what scientific reasoning is supposed to begin with. We start (or else it is no use starting) with an exact apprehension of the facts of the case, with a reliable transcript of the evidence of the senses which inductive reasoning can thereupon compound into more general truths or natural laws. We are led into error (according to inductive theory) only when the facts we thought we could rely upon were wrongly apprehended. Error is due to an indistinctness of vision, a false reading of that Book of Nature in which the truth resides and can be got at if only we can retain or reacquire the innocent, candid, childlike faculty of grasping what is in fact the case.

I share Karl Popper's view[1] that this conception of truth and error is utterly unrealistic. Scientific theories (I have said) begin as imaginative constructions. They begin, if you like, as stories, and the purpose of the critical or rectifying episode in scientific reasoning is precisely to find out whether or not these stories are stories about real life. Literal or empiric truthfulness is not therefore the starting-point of scientific enquiry, but rather the direction in which scientific reasoning moves. If this is a fair statement, it follows that scientific and poetic or imaginative accounts of the world are not distinguishable in their origins. They start in parallel, but diverge from one another at some later stage. We all tell stories, but the stories differ in the purposes we expect them to fulfil and in the kinds of evaluations to which they are exposed.

The divergence of poetic from factual truthfulness was not always taken for granted. For Sir Philip Sidney and his contemporaries it was something that had to be justified and reasoned out. 'Now for the poet,' says Sir Philip Sidney in a famous passage of his *Apology*, 'he nothing affirms and therefore never lieth. For, as I take it, to lie is to affirm that to be true which is false . . . but the poet (as I said before) never affirmeth.' If, all other things being equal, the choice is between correspondence with and departure from reality, then the choice is for reality: a

[1] 'On the Sources of Knowledge and of Ignorance', reprinted in *Conjectures and Refutations* (London, 1963).

painting which professes to be a portrait must be a likeness; but, if the choice is between what things are and what they ought to be, 'considered in relation to use and learning', then the literal truth, what actually happened, is usually less *doctrinable* than things as they might have been. For the scientist (Sidney says 'historian', but in this context scientist will do) is in bondage to the particular, to that which was – the historian's 'bare *was*' is Sidney's phrase – and any precept or general statement compounded of these bare particulars can only have the force of a 'conjectured likelihood'. It will not have the force of a poetic truth.

The idea that a poetic truth is a revelation of the ideal, of what *ought to be*, is taken by Sidney from Aristotle. Sidney (and incidentally Bacon) construe *ought to be* in the moralistic or doctrinary sense. For Bacon[1] narrative poetry 'feigns acts more heroical' than anything which actually happened, and thus 'conduces not only to delight but also to magnanimity and morality'. Dramatic poetry may be 'a means of educating men's minds to virtue', and the purpose of what Bacon describes as the highest form of poetry, the parabolical, must obviously be to improve.

This is what Sidney understood by the concept of what *ought to be*,[2] but according to Butcher's well-known analysis of the matter, it was not Aristotle's. The reason why Aristotle believed poetry to be 'a more philosophical and a higher thing than history' (and here, too, we may read 'science') is because it reveals what ought to be in the light of a true understanding of nature's intentions – not of nature's actions, for these are clumsy and imperfect. No, the poet discerns the purpose which nature is working, often most imperfectly, to fulfil. The poet is thus one up on nature (this was not Butcher's expression) and is the spokesman of her unfulfilled designs.

Aristotle's conception enriches or replaces scientific truth by truth of a higher kind, that which represents the testimony of a deeper and more privileged insight – a truth so lofty that, if nature does not conform to it, why then, so much the worse for nature.

[1] Francis Bacon, *De augmentis scientiarum*, book 2, chapter 13.
[2] See S. H. Butcher, *Aristotle's Theory of Poetry and Fine Art*, 1st ed. (1894); see also Ingram Bywater, *Aristotle on the Art of Poetry* (Oxford, 1909), and D. S. Margoliouth, *The Poetics of Aristotle* (London, 1911). 'What ought to be' is so rendered by Butcher and Bywater; Margoliouth writes 'the ideal'.

A second interpretation of poetic truth – the one I have just outlined is no longer professionally defended, which is not to say that it is no longer believed in – would claim for it that it represents truth not of a higher kind, but simply of a different kind, an alternative conception, or one of a set of alternatives, which enriches our understanding of the actual by making us move and think and orientate ourselves in 'a domain wider than the actual'. I believe this view is essentially a fair one, and it would be silly to squabble over matters of copyright to do with the usage of the word 'truth'. Nevertheless, great difficulties arise when it is allowed to infiltrate into science.

In this second conception of truth, a structure of imaginative thought – for example, a myth, especially if it appeals to magical agencies – will be judged true if it is all of a piece, hangs together, doesn't contradict itself, leaves no loose ends, and can cope with the unexpected. No single word in common speech describes this set of properties, but a narrative or theory or world picture or imaginative structure of any kind which answers to them is said to 'make sense', to have the property of being *believable-in*. All scientific theories must make sense, of course, but in addition they are expected to conform to reality, to be empirically true. It is the relaxation of this condition, or the failure to enforce it, that opens up to us a world that is larger, more various, and perhaps more doctrinable than real life.

I spoke of myths. In his famous work on savage thought, C. Lévi-Strauss[1] dismisses the cosy traditional belief that myths are primitive absurdities, are silly, innocent constructions that represent a merely rudimentary stage in the development of scientific thought. On the contrary, one can think of 'the rigorous precision of magical thought and ritual practices as an expression of the unconscious apprehension of the *truth of determinism*, the mode in which scientific phenomena exist'. Instead of contrasting magic and science, 'it is better to compare them as two parallel methods of acquiring knowledge', or as 'two scientific levels at which nature is accessible to scientific enquiry', both being 'equally valid'.

What Lévi-Strauss is telling us is that myths make sense, as conventional scientific theories make sense, and he does not feel that their failure to measure up to reality – to pass that extra examination which

[1] *La Pensée sauvage* (Paris, 1962), trans. as *The Savage Mind* (London, 1966).

has to do with conformity to real life – disqualifies them from being described as 'scientific'. Some Siberian peoples, he tells us, 'believe that the touch of a woodpecker's beak will cure toothache', and for this and similar reasons

it may be objected that science of this kind can scarcely be of much practical effect. The answer to this is that its main purpose is not a practical one. It meets intellectual requirements rather than or instead of satisfying needs . . . The real question is not whether the touch of a woodpecker's beak does in fact cure toothache. It is rather whether there is a point of view from which a wood-pecker's beak and a man's tooth can be seen as 'going together' (the use of this congruity for therapeutic purposes being only one of its possible uses), and whether some initial order can be introduced into the universe through these groupings.

This is a clear statement of his case, and I find it utterly unconvincing. Whose 'intellectual requirements' are being met, we may wonder, the savage's or the anthropologist's? By what extra criterion shall we be satisfied that the anthropologist himself is not creating a metamyth-ology, a mythology about myths? And would not someone actually suffering from toothache incline toward a more pragmatic style of thought? The point is that making sense and being believable-in are necessary but not sufficient qualifications for a process of intellection to be called commonsensical or scientific. The world of myths is Blake's world, Beulah, 'a place where contrarieties are equally true',[1] a world where the opposite of truth is not falsehood, but another truth; not necessarily a rival truth, but the telling of a different story, the testimony of a different interpretation of the world. Another myth, another set of magical allegiances, may serve the same or an equivalent purpose. The evidence Lévi-Strauss brings forward to contest the commonplace view that myths are a kind of fumbling approximation to science – a first groping attempt to make sense out of the complexi-ties of the world – is just that which seems to me to justify it. For myths are not really truths; at best they are truthlike structures, a part of the candidature for what might pass as true, but a candidature ex-cused from public examination.

The insufficiency of merely making sense and conferring order is not always fully grasped by laymen. Freudian psychoanalytic theory is a

[1] William Blake, *Milton*, book 2, plate 30.

mythology that answers pretty well to Lévi-Strauss's descriptions. It brings some kind of order into incoherence; it, too, hangs together, makes sense, leaves no loose ends, and is never (but never) at a loss for explanation. In a state of bewilderment it may therefore bring comfort and relief. But what about its therapeutic pretensions? The embarrassment of the woodpecker's beak is now got out of most adroitly. For in the opinion of many advanced thinkers, it is rather – well, rather *common* to suppose that the purpose of Freudian psychotherapy is, in the conventional sense, to cure. Its purpose is rather to give its subject a new and deeper understanding of his own condition and of the nature of his relationship to his fellow men. A mythical structure will be built up around him which makes sense and is believable-in, regardless of whether or not it is true. Another such structure might do as much for him – or as little.

In existential psychiatry,[1] the idea of 'cure' is dismissed contemptuously and replaced by the idea of 'healing'. A madman, for example, is healed when a microcosm of thought and personal relationships is built up around him in which his behaviour is no longer 'mad', that is, incongruous, anti-social, alienated from the majority opinion. The concept of 'explanation' is replaced by that of *understanding*, the process of discernment that uncovers a scheme of thinking within which a madman's actions and opinions now make sense.

With Freudian or existential psychology, as with myths, the question hardly arises of rational agreement or disagreement: these are ugly, hectoring words. Rather it is a question of acquiescence, of being taken into the author's scheme of thinking – and to describe acquiescence as a process of being 'taken in' has exactly the right connotation of surrender on the one hand and on the other hand of magic or contrived illusion. For these well-intentioned people are *telling stories*, sometimes wonderfully imaginative stories and sometimes wonderfully well told, so perhaps we should exercise a grown-up indulgence. When children don't tell the truth, their mother doesn't summon them to her knees and call them flaming little liars. On the contrary, she says, 'You

[1] My authorities here are mainly D. Cooper, *Psychiatry and Anti-Psychiatry* (London, 1967), and R. D. Laing, *The Self and Others* (London, 1961) and *The Divided Self* (London, 1960); and the occasional writings of both. See also M. Foucault, *Madness and Civilization*, English trans. (London, 1967), and J. Lacan, *Écrits* (Paris, 1966).

mustn't tell stories'; but although she wears a special kind of solemn face when she says so, she doesn't really think that telling stories is wicked unless it actually leads to harm.

Unfortunately, the psychologies I have been talking about are highly mischievous, not so much because they do harm or fail to do good, but because they represent a style of thought that will impede the growth of our understanding of mental illness. Consider for a moment imbecility, a subject in which scientifically founded psychiatry has made some ground. Here is an imbecile child which when it was born seemed ordinary. What can be wrong? Did some immemorial foreknowledge of the intrinsic contradictions of living drive it back into the habitation of a voiceless inner world? Did its parents, by some involuntary withholding of compassion, fail to ratify the child's ontological awareness of its essential self? Or is it perhaps unable to metabolise phenylalanine? Does it have the right number of chromosomes? What about the concentration of triiodothyronine in its blood? Two quite different sets of questions, and the people who ask them belong to two quite different kinds of worlds, figuratively speaking, the salon and the laboratory. Cultural psychiatry (but here I exclude Freud himself) repudiates the idea of an organic cause of mental abnormality; repudiates, indeed, some of the very concepts in terms of which the notion is expressed. The scientist *wants* it to be true. If there existed in science and medicine an analogue of literary criticism, we should investigate not only what people have reason to believe in, but the kind of things they *want* to believe in, and the cultural history of how they have come to acquire two or more different habits of expectation which cannot be reconciled.

It follows from what I have said that Freudian and other quasi-scientific psychologies are getting away with a concept of truthfulness which belongs essentially to imaginative literature, that in which the opposite of a truth is not falsehood but (we are back in Beulah) another truth. I strongly suspect that the same may be true of the more literary forms of other behavioural sciences, but I have not studied them deeply enough to say so for sure.

5

My contention has been that science tends to expel literature, and literature science, from any territory to which they both have claims – particularly the areas of learning that relate to human behaviour in its widest sense.

The distinguishing marks of the literary syndrome in science are, if you have followed my argument, these. *First*, there is an open or implied claim to a higher insight than can be achieved by laboratory scientists or historians or philologists, or by philosophers of the traditional English kind, an insight which soars beyond the busy little world of test tubes and graphs and measuring instruments, or indeed of facts. *Second*, there is a combination of high imaginativeness with a relaxation of or a failure to enforce the critical process, so that the critical and inventive faculties no longer work together synergistically, but tend if anything to compete; and with this goes a whispering campaign against the importance attached to validation or justification and even, in extreme cases, now beyond remedy, against rational thought. *Third* (and this is what gives the syndrome a literary rather than a metaphysical character) is the *style* in which the high truths of the imagination are made known, a style which (among many other disfigurements) deliberately exploits the voluptuary and rhetorical uses of obscurity, a style which at first intrigues and dazzles, but in the end bewilders and disgusts.

One may well ask: If the forms of discourse that answer to this description are kept outside the reach of a critical apparatus; if in repudiating the ideas of proof or cure or any other scheme of validation they escape the sanctions that are enforced upon physicians or historians or laboratory scientists, what then is to stop them from expanding their influence and pretensions without limit? The answer is clear enough. They are not repudiated, but as fashion changes they will be forgotten, to be classified as a scientific curiosity or literary genre, as dead as the Philosophic Romances of the seventeenth century or the System Philosophies of the nineteenth. This fate is the unhappiest that could befall them, because their practitioners want above all else to be in the swim, to be counted among the makers of cultivated opinion, rare spirits whose thought transcends the busy preoccupations of common people. To be forgotten is the worst of their bad dreams.

There is an aberration of science or of the scientific style of thinking which has come to be known as 'scientism'. Roughly speaking, it stands for the belief that science knows or will soon know all the answers, and it has about it the corrupting smugness of any system of opinions which contains its own antidote to disbelief. I suppose my lecture has been about *poetism*, an aberration of imaginative literature about which (*mutatis mutandis*) one could say very much the same. It stands for the belief that imaginative insight and a mysteriously privileged sensibility can tell us all the answers that are truly worthy of being sought or being known, and its practitioners are rallied by the inane war cry that beauty is equivalent to truth.

For scientism, imaginative literature is best thought of as a branch of the entertainment industry; for poetism, scientists are engaged in merely parsing the Book of Nature, the inner meaning of which they are altogether unqualified to comprehend. Poetism is only a minor ailment of literature, but an ailment that literature is prone to through an excess of its own exuberant strength; in the time scale of literature, its outbreak seems to be a seasonal event. Scientism, for just the same reason, is latent in scientific thinking – a malady to which, because of our constitutions, we scientists are specially predisposed. Both views are about equally contemptible, so there is no need for us to take sides. No one need beat his breast and say, 'Now *I* am on the side of the poets', because poets are not really on that side and scientists not really on the other. I admit that it was mainly a love of science that prompted me to speak as I have done, but it could equally well have been a love of literature, and if it had been so I do not think that my lecture would have been so very different from that which you have just heard.

POSTSCRIPT

The omission of C. P. Snow's name and any discussion of the arguments of his famous Rede Lecture was deliberate. The trouble is that upon mention of Snow's name literary intellectuals spring at once into ungainly postures which make rational argument almost impossible. Some such intellectuals behave as if Snow had either been responsible for, or at least advocated, the existence of two cultures. In reality he was calling attention to the fact that educated and reflective people subsist upon diets of two very different kinds: one having to do with scientific theories and ideas, the other literary and more overtly imaginative in character. This always struck me as being a straightforwardly objective observation and as objectively true as it is to say that Italians have a special liking for pasta and Scottish people for oatmeal in all its allotropic forms. Such a statement embodies no expression of opinion upon the superiority of either. It is their unlikeness and not their inequality of merit that is in question.

The curse of it is that people *will* take sides. All those whose schooldays were darkened by profitless and apparently interminable discussions or formal debates on the general theme of 'The sciences v. the humanities' will share my dismay that so many of my critics chose to regard my Romanes Lecture as yet another contribution to this idiotic debate. It was nothing of the kind. I expressed, and I take this opportunity to reaffirm, my abhorrence equally of 'scientism' and 'poetism' (p. 60). Scientism deserves no friends and has been so skilfully attacked that it needs no further enemies; but there is some recent writing in 'cultural' anthropology and psychiatry which convinces me that science is in real need of some defence against poetism. Poetism is the undisciplined exercise of the imaginative faculty to produce hypotheses which are held to be true, and defended against all criticism, merely because of their supposedly high inspirational origin or because they are specially well put or make an unusually strong appeal to some dark visceral mystic predilection of their authors. The hero of poetism is, of course, that William Blake who 'came in the grandeur of inspiration to cast off rational demonstration'.[1]

[1] Blake, *Milton*, book 2.

FURTHER COMMENTS ON
PSYCHOANALYSIS

IN my Romanes Lecture on science and literature I implied that a psychoanalytical explanation-structure answered pretty closely to Lévi-Strauss's description of a myth. By this I meant that a psychoanalytical interpretation weaves around the patient a well-tailored personal myth within the plot of which the subject's thoughts and behaviour seem only natural, and, indeed, only what is to be expected.

I must begin by making it clear that my criticism of psychoanalysis is not to be construed as a criticism of psychiatry or psychological medicine as a whole. People nowadays tend to use 'psychoanalysis' to stand for all forms of psychotherapy, much as 'Hoover' is used as a generic name for all vacuum cleaners and 'Vaseline' for all ointments of a similar kind. By psychoanalysis I understand that special pedigree of psychological doctrine and treatment which can be traced back, directly or indirectly, to the writings and work of Sigmund Freud. The position of psychological medicine today is in some ways analogous to that of physical or conventional medicine in the middle of the nineteenth century. The physician of a hundred and thirty years ago was confronted by all manner of medical distress. He studied and tried to cure his patients with great human sympathy and understanding and with highly developed clinical skills, by which I mean that he had developed to a specially high degree that form of heightened sensibility which made it possible for him to read a meaning into tiny clinical signals which a layman or a beginner would have passed over or mis-understood. The physician's relationship to his patient was a very personal one, as if healing were not so much a matter of applying treatment to a 'case' as a collaboration between the physician's guidance and his patient's willingness to respond to it. But – there was so little he could do! The microbial theory of infectious disease had not been formulated, viruses were not recognised, hormones were unheard of, vitamins undefined, physiology was rudimentary and biochemistry almost non-existent.

The psychiatry of today is in a rather similar position, because we are still so very ignorant of the mind. But the best of its practitioners are people of great skill and understanding and apparently inexhaustible patience; people whose humanity reveals itself just as much in the way they recognise their limitations as in their satisfaction when a patient gets better in their care. I am emphasising this point to make it clear that to express dissatisfaction with psychoanalysis is not to disparage psychological medicine as a whole.

One of my critics has accused me of saying or implying that he, a psychoanalyst, would attempt to treat by psychiatric means the symptoms of a brain tumour or of Huntington's Chorea. *Of course* I don't think a psychoanalyst would knowingly attempt to treat a brain tumour or a victim of Huntington's Chorea by psychoanalytic methods, but he may not realise the degree to which he is being wise after the event. Being a sensible man he naturally repudiates the idea of treating those psychological ailments of which physical causes are, in general terms, already known. But psychoanalysts do treat and speculate upon the origins of schizophrenic conditions and manic-depressive psychoses. *These* are the test cases: what are we to make of *them*?

Are 'mental illnesses' of mental or physical origin? To answer this question I shall begin with what may appear to be a digression. As recently as thirty years ago, many geneticists were still worried and confused by the problem of assessing, in precise terms, the relative contributions of nature and nurture – of heredity and environment or upbringing – to the overt ('phenotypic') differences between our mental and physical constitutions and capabilities. Both nature and nurture exercise an influence, of course; but L. T. Hogben and J. B. S. Haldane were the first to make it publicly clear that there is no *general* solution of the problem of estimating the size of the contribution made by each. The reason is that the size of the contribution made by nature is itself a function of nurture. (I use the word 'function' in its mathematical sense.)[1] If someone constitutionally lacks the ability to synthesise an essential dietary substance, say X, then the contribution made by heredity to the difference between himself and his fellow men will depend on the environment in which they live. If X is abundant in the

---

[1] In mathematics, x is a function of y when the value of x varies in dependence on the value of y.

food he normally has access to, his inborn disability will put him at no disadvantage and may not be recognised at all; but if X is in short supply or lacking, then he will become ill or die. The same reasoning applies to other, much more complicated examples. If people live a simple pastoral life that makes little demand on their resourcefulness and ingenuity, inherited differences of intellectual capability may not make much difference to their behaviour; but it is far otherwise if they live a difficult and intellectually demanding life. How often has it not been said that the stress of modern living raises the threshold of competence below which people can no longer keep up or make the grade? This is not to deny that some differences between us are for all practical purposes wholly genetic, wholly inborn. A person's blood group is described as 'inborn' not just because it is specified by his genetic make-up, but because (with certain rare and known exceptions) there is no environment capable of supporting life in which that specification will not be carried out. Most differences between us are determined both by nature and by nurture, and their contributions are not fixed, but vary in dependence on each other.[1]

With this analogy in mind, let me now turn to psychological disorders, which – to beg no questions – I shall define as conditions which cause a person to seek, or need, or be directed towards, the care of a psychiatrist. Here, too, as a first approximation, it will be reasonable to assume that both 'mental' and 'organic' states or agencies contribute to the difference between the psychiatrist's patient and his fellow men; but here, too, we should be very cautious in our attempts to assign precise values to the contributions made by each. It seems natural to repudiate the idea of psychiatric treatment of brain tumours, because they seem so obviously organic in origin; but even in this extreme case we mustn't be too sure. Many of us now believe that there exists a natural defensive mechanism against tumours which is of the same general kind as that which prohibits the transplantation of tissues from one individual to another. If these natural defences are indeed immunological in nature, they are open to influences of a kind that common

---

[1] To speak (as I do here and below) of the causes of *differences* between human beings sounds clumsy and takes some getting used to; but there seems to be no avoiding it if one is to be precise and at the same time avoid a formal symbolic treatment.

sense will classify as mental, or anyhow behavioural, e.g. to prolonged frustration, unhappiness, distress, or indifference to living.[1] (The psychosomatic element in tuberculosis is specially relevant here, because the natural defence against tuberculosis depends on immunological mechanisms of a very similar kind.)

To go now to the other extreme: the psychoanalytic critic I referred to above thinks it probable that 'neurosis is the result of faulty early conditioning' rather than of brain disease or an inborn error of metabolism. No doubt; but does he not also think that constitutional or organic influences may raise or lower the susceptibility of his patients to these disturbing influences? Of course he does – and so did Freud. It is normally a mistake, I suggest, to trace any psychological disorder to wholly mental or wholly organic causes. Both contribute, though sometimes to very unequal degrees, and the contribution made by one will be a function of the contribution made by the other.

It is, nevertheless, very understandable that psychiatrists should approach their patients with two rather different kinds of aetiological purpose and interest in mind. Psychiatrist A will say, 'My interest lies in trying to see how a certain pattern of upbringing, environment, habits of life and human relationships may predispose people of certain constitutions to psychological disorders.' Psychiatrist B will say, 'Now *my* interest lies in trying to identify those elements of heredity and organic constitution which make a man specially likely to contract a certain psychological disorder if he is influenced by the environment and his fellow men in certain ways.' Both attitudes seem very reasonable, and over much of the territory that belongs to them the two psychiatrists will not compete. But – and now I come to my main point – in the context of those serious psychological disorders that are still disputed territory, the methodology implicit in the attitude of Psychiatrist B is very much the more powerful.

The reason is this. A physical abnormality can be the subject of diagnosis, and therefore, in principle, of treatment or avoiding action, *before* it can contribute to a psychological disturbance. The recognition early in life of a certain physical abnormality (say, the chromosomal constitution XYY) defines a priori a category of men who are at special risk; and our foreknowledge of that risk can be made the basis

[1] See 'The Crab', pp. 154–66 below, especially pp. 161–5.

of a rational system of avoidance. The physical disability represents a parameter of the situation, where upbringing and environment are variables which can be varied within certain limits at our discretion. A difficult enterprise, to be sure; but not so difficult as, and much more realistic than, say, abolishing all family life, as one 'existential psychiatrist' is alleged to have recommended, because some families may create an environment conducive to mental disorder. With certain forms of low-grade mental deficiency, this programme is now adopted as a matter of routine. When tests carried out on a baby's urine suggest that it cannot metabolise the amino-acid phenylalanine, its diet can be altered in such a way as to prevent or reduce the severity of what might otherwise be irremediable damage to the brain. I hope and expect that cognate solutions will one day be found for the major psychoses. No matter what other factors may have influenced him, there is something organically wrong with a manic-depressive patient, and it is essential to find out what it is, preferably before he becomes gravely ill.

This completes my attempt to explain why I think that the categorical distinction between brain disease and mental illness, as between 'Nature' and 'Nurture', is a fundamentally unsound one – the remnant of an effete dualism, a still further perpetuation of what Ryle called the legend of Two Worlds.

I now turn to psychoanalysis itself, taken in the sense I gave it in an earlier paragraph. I shall not attempt a systematic treatment, but shall merely draw attention to a few of its more serious methodological, doctrinal and practical defects.

The property that gives psychoanalysis the character of a mythology is its combination of conceptual barrenness with an enormous facility in explanation. To criticise a theory because it explains everything it is called on to explain sounds paradoxical, but anyone who thinks so should consult the discussion by Karl Popper in *Conjectures and Refutations*, particularly the passages that make mention of psychoanalysis itself.[1] Let me illustrate the point by a number of passages chosen from the authors' summaries of their own contributions to the 23rd International Psychoanalytical Congress held in Stockholm in 1963. I choose the proceedings of a congress rather than the work of a single author so as to get a cross-section of psychoanalytic thought.

[1] *Conjectures and Refutations* (London, 1963), pp. 34–9.

Character-traits are formed as precipitates of mental processes. They originate in innate properties; they come into existence in the mutual interplay of ego, id, super-ego and ego-ideal, under the influence of object-relations and environment.

When an individual strikes out at his wife, his child, his acquaintances or even complete strangers, we may well suspect that a gross failure in Ego-functioning has occurred. Its restraining control has been partially eluded.

Of a 'cyclothymic' patient in the fifth and sixth years of psychoanalytic treatment:

. . . the delusion of having black and frightening eyes took the centre of the analytic stage following the resolution of some of the patient's oral-sadistic conflicts. It proved to be a symptom of voyeuristic tendencies in a split-off masculine infantile part of the self and yielded slowly to reintegration of this part, passing through phases of staring, looking at and admiring the beauty of women.

On the aetiology of anti-Semitism:

The Oedipus complex is acted out and experienced by the anti-Semite as a narcissistic injury, and he projects this injury upon the Jew who is made to play the role of the father . . . His choice of the Jew is determined by the fact that the Jew is in the unique position of representing at the same time the all-powerful father and the father castrated . . .

On the role of snakes in the dreams and fantasies of a sufferer from ulcerative colitis:[1]

The snake represented the powerful and dangerous (strangling), poisonous (impregnating) penis of his father and his own (in its anal-sadistic aspects). At the same time, it represented the destructive, devouring vagina . . . The snake also represented the patient himself in both aspects as the male and female and served as a substitute for people of both sexes. On the oral and anal levels the snake represented the patient as a digesting (pregnant) gut with a devouring mouth and expelling anus . . .

I have not chosen these examples to poke fun at them, ridiculous though I believe them to be, but simply to illustrate the Olympian glibness of psychoanalytic thought. The contributors to this congress

[1] A disease of the kind psychoanalysts would be well advised not to meddle with.

were concerned with homosexuality, anti-Semitism, depression, and manic and schizoid tendencies; with *difficult* problems, then – problems far less easy to grapple with or make sense of than anything that confronts us in the laboratory. But where shall we find the evidence of hesitancy or bewilderment, the avowals of sheer ignorance, the sense of groping and incompleteness that is commonplace in an international congress of, say, physiologists or biochemists? A lava-flow of *ad hoc* explanation pours over and around all difficulties, leaving only a few smoothly rounded prominences to mark where they might have lain. Surely the application of psychoanalytic methods in a completely alien culture might give even the most sanguine practitioner reason to pause? Not a bit of it. We have the word of two of the contributors to the congress that 'the usual technique and theory of psychoanalysis were found to be applicable to obtain an understanding of the inner life' of the Dogon peoples in Mali:

A twenty-four-year-old Dogon man, who at the beginning had met the white stranger with profound distrust, was led to change his views with surprising speed.

After first having built a subsidiary transference and involved a younger colleague in the analysis, he turned from the animate object to the inanimate (playing with sticks) and from this to tactile gestures . . . Finally he 'regressed' to somatic forms of expression in that he continued the analytic exchange by urinating . . .

The examples I have chosen above, and the psychoanalytic autopsies I shall mention later, illustrate another important methodological defect of psychoanalytic theory. If an explanation or interpretation of a phenomenon or state of affairs is to be fully satisfying and actable-on, it must have a special, not merely a general relevance to the problem under investigation. It must be, rather specially, an explanation of whatever it is we want to explain, and not also an explanation of a great many other, perhaps irrelevant things as well.

For example: if a patient cannot retain salt in his body, it is not good enough (though it will probably not be wrong) to say that his endocrine system is in disorder, because such an explanation would cover a multitude of other abnormalities besides. The explanation may well be that the patient is no longer producing aldosterone, a specific hormone of the cortex of the adrenal gland, and if that is so he can probably be cured. Again, it will not do to say that muscular contraction is a transformation of energy derived originally from the sun. This is a weak explanation; it is too far removed in the pedigree of causes; we are more interested in the causal parentage of the phenomenon than in its causal ancestry. Strong explanations have a quality of *special* relevance, of logical immediacy: and this is a quality they must have if they are to be tested and shown to be acceptable for the time being or, as the case may be, unsound. Psychoanalytic explanations are invariably weak explanations in just this sense.

'Validation of psychoanalytic theory is a difficult business', my psychoanalytic disputant said, though he betrayed no logical understanding of why it should be so; and by implication he suggested that, instead of criticising it destructively, I should help find means of testing whether or not it is true. Alas – except in one respect, which I shall deal with in a moment – the methodological obstacles are insuperable. Indeed, psychoanalysis has now achieved a complete intellectual closure: it explains even why some people disbelieve in it. But this accomplishment is self-defeating, for in explaining why some people do not believe in it, it has deprived itself of the power to explain why other people do. The ideas of psychoanalysis cannot both be an object of critical scrutiny and at the same time provide the conceptual background of the method by which that scrutiny is carried out.

It is for this reason that the notion of *cure* is methodologically so important. It provides the only independent criterion by which the acceptability of psychoanalytic notions can be judged. This is why cure is such an embarrassment for 'cultural' psychiatry in general. No wonder its practitioners try to talk us out of it,[1] no wonder they prefer to see themselves as the agents of some altogether more genteel ambition, for example, to give the patient a new insight through a new deep, inner understanding of himself. But let us not be put off. Some people get better *under* psychoanalytic treatment, of course; but do they get

[1] 'Curing is so ambiguous a term', says Dr David Cooper in *Psychiatry and Anti-Psychiatry*; 'one may cure bacon, hides, rubber, or patients. Curing usually implies the chemical treatment of raw materials so that they may taste better, be more useful, or last longer. Curing is essentially a mechanistic perversion of medical ideals that is quite opposite in many ways to the authentic tradition of healing.' Somewhat similar views are to be found in the writings of R. D. Laing, Michel Foucault and J. Lacan.

better as a specific consequence of psychoanalysis as such? Consider an example. A young man full of anxieties and worries seeks treatment from a psychoanalyst, and after eighteen months' or two years' treatment finds himself much improved. Was psychoanalytic treatment responsible for the cure? One cannot give a confident answer unless one has reasonable grounds for thinking:

(*a*) that the patient would not have got better anyway;

(*b*) that a treatment based on quite different or even incompatible theoretical principles, for example, the theories of a rival school of psychotherapists, would not have been equally effective; and

(*c*) that the cure was not a by-product of the treatment. The assurance of a regular sympathetic hearing, the feeling that somebody is taking his condition seriously, the discovery that others are in the same predicament, the comfort of learning that his condition is explicable (which does not depend on the explanation's being the right one) – these factors are common to most forms of psychological treatment, and the good they do must not be credited to any one of them in particular. At present there is no convincing evidence that psychoanalytic treatment as such is efficacious, and unless strenuous efforts are made to seek it the entire scheme of treatment will degenerate into a therapeutic pastime for an age of leisure.

The lack of good evidence of the specific therapeutic effectiveness of psychoanalysis is one of the reasons why it has not been received into the general body of medical practice. A layman might be inclined to say that we should give it time, for doctors are conservative people and ideas so new take ages to sink in. But it is only on a literary time-scale that Freudian ideas are new. By the standards of current medical practice they have an almost antiquarian flavour. Many of Freud's principles were formulated before the recognition of inborn errors of metabolism, before the chromosomal theory of inheritance, before even the rediscovery of Mendel's laws. Hormones were unheard of when Freud began to propound his doctrines, and the mechanism of the nervous impulse, of which we now have a pretty complete understanding, was quite unknown.

Nevertheless, psychoanalysts are wont to say that Freud's work carried conviction because it was so firmly grounded on basic biological principles. I am therefore sorry to have to express the professional opinion that many of the germinal ideas of psychoanalysis are profoundly unbiological, among them the 'death-wish', the underlying assumption of an extreme fragility of the mind, the systematic depreciation of the genetic contribution to human diversity, and the interpretation of dreams as 'one member of a class of *abnormal* psychical phenomena'.

I said earlier that the mythological status of psychoanalytic theory revealed itself in its combination of unbridled explanatory facility with conceptual barrenness, a property to which I have not yet referred. Ever since Freud's factually erroneous analysis of Leonardo,[1] psychoanalysts have tried their hand at 'interpreting' the life and work of men of genius, and many of the great figures of history have been disinterred and brought to the post-mortem slab. The fiasco of Darwin's retrospective psychoanalysis has already been held up to ridicule.[2] But, Darwin apart, how can we not marvel at the way in which the whole exuberant variety of human genius can be explained by the manipulation of a handful of germinal ideas – the Oedipus complex, the puzzlement of discovering that not everyone has a penis, a few unspecified sado-masochistic reveries, and so on: surely we need a more powerful armoury than this? Evidently we do, for these analyses always stop short of explaining why genius took the specific form that interests us. Freud does not profess to tell us why Leonardo became an artist. 'Just here our capacities fail us', he says, with a modesty not found in the writings of his successors; but it is hard not to feel let down.

A critique of psychoanalysis is, in the outcome, never much more than a skirmish, because (as I tried to explain) its doctrines are so cunningly insulated from the salutary rigours of disbelief. It is nevertheless customary to end any such critique with a spaciously worded acknowledgement of our indebtedness to Freud himself. We recognise his enlargement of the sensibilities of physicians, his having opened up a new era of human speculation, his freeing us from the confinements of prudery and self-righteousness, etc. There is some truth in all of this. There is some truth in psychoanalysis too, as there was in Mesmerism and in phrenology (for example, the concept of localisation of function in the brain). But, considered in its entirety, psychoanalysis

---

[1] See B. A. Farrell, introduction to *Leonardo da Vinci* (Harmondsworth, 1963).
[2] 'Darwin's Illness', pp. 141–7 below, especially pp. 142–4.

won't do. It is an end-product, moreover, like a dinosaur or a zeppelin; no better theory can ever be erected on its ruins, which will remain for ever one of the saddest and strangest of all landmarks in the history of twentieth-century thought.

POSTSCRIPT

My methodological criticisms of psychoanalysis deserve more attention than any psychoanalyst has yet found time or inclination to give them. While reaffirming my belief that these criticisms are valid let me again emphasise that they are expressly directed against psychoanalysis and not against psychiatry in general. This methodological criticism is of course far from complete. Professor Hugh Trevor-Roper (Lord Dacre) in the *Sunday Times* of 18 February 1973 has called attention to another methodological enormity – one specially perpetrated by 'psycho-historians'. The usual practice in science or historical research is to frame hypotheses in such a form that the facts follow from them, that is, in such a way that statements expressing the matters of fact in need of explanation are among the logical implications of the hypothesis. In psycho-history, however, the facts are shaped in such a way as to make them appear to follow from a preconceived hypothesis. This psycho-historical approach authorises us to declare with certainty that Hitler's character make-up and behaviour point to Mrs Hitler's extreme severity with young Adolf's toilet-training, a subject of which we are luckily quite ignorant.

INDUCTION AND INTUITION IN
SCIENTIFIC THOUGHT

I THE PROBLEM STATED

I

IT IS NOT at all usual for scientists to write about the nature of scientific method, particularly if they are still engaged in scientific research. It is however an understood thing that scientists of a specially elevated kind, theoretical physicists for example, may from time to time express quietly authoritative opinions on the conduct of scientific enquiry, while the rest of us listen in respectful silence; but that a biologist should speak up where so many physicists and chemists have chosen to remain silent must seem to be yet another symptom of the decay of values and the loss, in this modern world, of all sense of the fitness of things.

Yet – if the task of scientific methodology is to piece together an account of what scientists actually *do*, then the testimony of biologists should be heard with specially close attention. Biologists work very close to the frontier between bewilderment and understanding. Biology is complex, messy and richly various, like real life; it travels faster nowadays than physics or chemistry (which is just as well, because it has so much farther to go), and it travels nearer to the ground. It should therefore give us a specially direct and immediate insight into science in the making. The wisest judgements on scientific method ever made by a working scientist were indeed those of a great biologist, Claude Bernard.[1]

We all know in rough outline what lawyers do, or clergymen, physicians, accountants and civil servants; we have a vague idea of the codes of practice they must abide by if they are to succeed in their professional duties, and if we were to learn more about them we should be

---

[1] *Introduction à l'étude de la médecine expérimentale* (Paris, 1865), a work that has suffered in translation (which may account for its limited influence in the English-speaking world).

edified, no doubt, but not surprised. But what are scientists like as professional men, and how do they go about enlarging our understanding of the world around us? There seems to be no one answer. The layman's interpretation of scientific practice contains two elements which appear to be unrelated and all but impossible to reconcile. In the one conception the scientist is a discoverer, an innovator, an adventurer into the domain of what is not yet known or not yet understood. Such a man must be speculative, surely, at least in the sense of being able to envisage what *might* happen or what could be true. In the other conception the scientist is a critical man, a sceptic, hard to satisfy; a questioner of received beliefs. Scientists (in this second view) are men of facts and not of fancies, and science is antithetical to, perhaps even an antidote to, imaginative activity in all its forms.[1]

Let me begin with the scientist as a questioner of received beliefs. During the seventeenth century, when the new science came in on a spring tide,[2] and again during the nineteenth century, the forward movement of science called for a vigorous shaking off of scholastic constraints and religious superstition. No single work displays science in its critical temper more clearly then Francis Galton's *Statistical Inquiries into the Efficacy of Prayer*, published by the *Fortnightly Review* in its issue of 1 August 1872.

A belief in the efficacy of prayer (Galton reasoned) is something we all grow up with: it has behind it the formidable authority of habit, doctrine and popular assent. But are there any 'scientific' grounds for supposing that prayers are answered: that what is prayed for comes about as a consequence of an act of prayer? One line of enquiry that seemed to Galton to promise a definite answer turned upon the health and longevity of the Queen and other members of the royal family – something prayed for weekly or even daily on a national scale, and sung for too, though in an imperative rather than a supplicatory mood. Do members of royal families live any longer as a result of these exertions of prayer on their behalf? The table given here, based on one of Galton's, shows that if anything they fare worse than people of humbler birth.

---

[1] Cf. the views of Thomas Gradgrind, pp. 16–17 above.
[2] The phrase is Henry Power's: *Experimental Philosophy* (London, 1644; New York, 1966), p. 192.

Mean age attained by males of various classes who had survived their thirtieth year, from 1758 to 1843: deaths by accident or violence excluded

	Number of men	Average age at death	Eminent men[1]
Members of Royal houses	97	64.04	
Clergy	945	69.49	66.42
Lawyers	294	68.14	66.51
Medical profession	244	67.31	67.07
English aristocracy	1,179	67.31	
Gentry	1,632	70.22	
Trade and commerce	513	68.74	
Officers in the Royal Navy	366	68.40	
English literature and science	395	67.55	65.22
Officers of the Army	569	67.07	
Fine arts	239	65.96	64.74

[1] The eminent men are those whose lives are recorded in Alexander Chalmers's *General Biographical Dictionary* (32 vols, London, 1812–17), with some additions from the *Annual Register*.

The amplitude and frequency of prayers for the royal family cannot be assumed to be proportional to their sincerity, so Galton put the same question in a different way. No one can doubt the sincerity of prayers that appeal for the lives of newborn children: are then still-births any less frequent among the children of the devout than among the professional classes generally? Apparently not: Galton studied the number of stillbirths announced in the *Record* (a clerical newspaper) and in *The Times*, and found them to stand in exactly the same proportion to the total number of recorded deaths. The data are shaky, of course, and Galton was quite aware of the shortcomings of his analysis. His purpose was above all to show that such an analysis can be done.

Galton's most telling argument was founded upon the policy of insurance companies in fixing the rates of annuities. To buy an annuity is to pay a capital sum at (for example) retirement, in return for which the company undertakes to provide the investor with an annual income until he dies. The rates offered by different companies are competitive and must be judiciously worked out, for if the annuitant lives beyond the calculated expectation the insurance company will be out of pocket. This being so,

It would be most unwise, from a business point of view, to allow the devout, supposing their greater longevity even probable, to obtain annuities at the same low rates as the profane. Before insurance offices accept a life, they make confidential inquiries into the antecedents of the applicant. But such a question has never been heard of as, 'Does he habitually use family prayers and private devotions?' Insurance offices, so wakeful to sanatory influences, absolutely ignore prayer as one of them. The same is true for insurances of all descriptions, as those connected with fire, ships, lightning, hail, accidental death, and cattle sickness. How is it possible to explain why Quakers, who are most devout and most shrewd men of business, have ignored these considerations, except on the ground that they do not really believe in what they and others freely assert about the efficacy of prayer?

I have not done justice to the range and analytical skill of Galton's polished and urbane analysis, and I shall have done him a positive injustice if I leave you with the impression that he was merely having a go at religious belief. The rhetorical force of his argument would have been greatly weakened if it had been crudely irreligious. Prayer, he tells us, may strengthen the resolution and bring serenity in distress: it is an appeal for help; Galton did not 'profess to throw light on the question of how far it is possible for man to commune in his heart with God'. His reasoning was thus 'scientific' in the territory in which he exercised it, but also in the territory he disclaimed.

Reasoning in this style is by no means confined to or even specially characteristic of scientific enquiry. Galton's first step was to assume the truth of an opinion for which there was a certain obvious prima facie case, namely that what is prayed for may come about through prayer; then he examined some of the logical consequences of holding that opinion; then, thirdly, he took steps to find out whether or not those logical expectations were indeed fulfilled. The argument was made out by reasoning, not by asseveration; the matters of fact upon which the judgement turned were, if not known, then knowable by everyone; and the testimony of inner voices went unheard. His great achievement was, of course, methodological. He brought within the domain of science matters until then thought to lie outside its competence: 'the efficacy of prayer seems to me . . . a perfectly appropriate and legitimate subject of scientific inquiry'. The reasoning may be empirically wrong, but it is not fallacious; an answer will be arrived at by this style of reasoning or not at all.

The critical task of science is not complete and never will be, for it is the merest truism that we do not abandon mythologies and superstitions, but merely substitute new variants for old. No one of Galton's stature has conducted a statistical enquiry into the efficacy of psychoanalytic treatment. If such a thing were done, might it not show that the therapeutic pretensions of psychoanalysis were not borne out by what it actually achieved? It was perhaps a premonition of what the results of such an enquiry might be that has led modern psychoanalysts to dismiss as somewhat vulgar the idea that the chief purpose of psychoanalytic treatment is to effect a cure. No: its purpose is rather to give the patient a new and deeper understanding of himself and of the nature of his relationship to his fellow men. So interpreted, psychoanalysis is best thought of as a secular substitute for prayer. Like prayer, it is conducted in the form of a duologue, and like prayer (if prayer is to bring comfort and refreshment) it requires an act of personal surrender, though in this case to a professional and stipendiary god.

Nor has anyone yet conducted a formal analysis of the all but universal belief that dreams are messages of some kind; that dreams convey significant information clothed in a dark and ancient symbolism which only the initiated can decode. Analysis, I suspect, would reveal that dreams, whatever else they may be, are not messages or communications of any kind. The utter nonsensicality of dreams – their glorious emancipation from the confinements of time and place and cause and sense – is probably the most significant thing about them, the property from which the student of mind has most to learn. If these newer enquiries were to be set in train, and were to have the outcome I have predicted, the resentment and sense of outrage they would give rise to would be indistinguishable in character and psychological origin from that which exploded nearly one hundred years ago over Galton's analysis of prayer.

2

The layman's conception of the scientist as a critic, a sceptic, a man intolerant or contemptuous of conventional beliefs, is obviously incomplete. The exposure and castigation of error does not propel science forward, though it may clear a number of obstacles from its path. To prove that pigs cannot fly is not to devise a machine that does so. To explode the myth of the chimera makes it no easier to transplant a kidney from (say) ape to man.

The layman sees the other profile too. A scientist is a man who weighs the earth and ascertains the temperature of the sun; he destroys matter and invents new forms of matter, and one day he will invent new forms of life. But how has he achieved the understanding that makes this possible? What methods of enquiry apply with equal efficacy to atoms and stars and genes? What *is* 'The Scientific Method'? What goes on in the head when scientific discoveries are made?

Rhetorical questions: and when we try to answer them a remarkable state of affairs is revealed. The scholarly discipline that might be expected to hold the answers is unpopular and in the main, in its larger ambitions, unsuccessful. If the purpose of scientific methodology is to prescribe or expound a system of enquiry or even a code of practice for scientific behaviour, then scientists seem to be able to get on very well without it. Most scientists receive no tuition in scientific method, but those who have been instructed perform no better as scientists than those who have not. Of what other branch of learning can it be said that it gives its proficients no advantage; that it need not be taught or, if taught, need not be learned?

It will not do to say that a scientist learns by apprenticeship, implying that he learns to do his own work by studying the Works of others, for scientific 'papers' in the form in which they are communicated to learned journals are notorious for misrepresenting the processes of thought that led to whatever discoveries they describe. The scientist is not conscious of acting out a method. If a scientist is more or less successful in the enterprise he is engaged on, he attributes it to having enjoyed more or less of luck or learning or perceptiveness or flair, *never* to the use or misuse of a formal methodology. How very unlike a deductive exercise, such as we carry out when trying to derive a geometric theorem; here, if something goes wrong, our first thought is that we have made a logical (that is, a methodological) mistake.

Of course, the fact that scientists do not consciously practise a formal methodology is very poor evidence that no such methodology exists. It could be said – has been said – that there is a distinctive methodology of science which scientists practise unwittingly, like Molière's M.

Jourdain, who found that all his life, unknowingly, he had been speaking prose. Yet it may be revealing that not one of those whom we recognise as great methodologists of science was a practising scientist himself. Francis Bacon was a lawyer and a man of affairs; a sociologist of science, if you like, and (if you like) a playwright. John Stuart Mill was a deeply learned and humane man, a political theorist and a sociologist in the modern sense, but though his 'strong relish for accurate classification' had been gratified by lectures and books on botany and zoology,[1] his deeper scientific knowledge came at second hand from William Whewell's *History of the Inductive Sciences* (1837). Whewell himself did not practise science nor add to it, except by way of nomenclature, but he was deeply enough informed about all its branches to have become the outstanding methodologist of his day.[2] Karl Pearson was a mathematician; Stanley Jevons and John Maynard Keynes were economists; C. S. Peirce was, as Karl Popper is, a great philosopher. Why did not scientists come forward and expound their own methodology? One did so: Claude Bernard, whom I have already mentioned; but his opinions seem to have made so little impact on the English-speaking world that his name is mentioned in only two of a dozen well-known texts on scientific methodology on my shelves.

Unfortunately, a scientist's account of his own intellectual procedures is often untrustworthy. 'If you want to find out anything from the theoretical physicists about the methods they use,' said Albert Einstein, 'I advise you to stick closely to one principle: don't listen to

[1] J. S. Mill, *Autobiography* (London, 1873). Mill attended lectures on zoology in Montpellier in 1820, and there seems no doubt that his thought on methodology was strongly influenced by the study of a subject overwhelmed by a multitude of 'facts' which had not yet been disciplined by a unifying theory. Coleridge described it as 'notorious' that zoology had been 'falling abroad, weighed down and crushed as it were by the inordinate number and multiplicity of facts and phenomena apparently separate, without evincing the least promise of systematising itself by any inward combination of its parts'. *General Introduction*, or *A Preliminary Treatise on Method: Encyclopaedia Metropolitana* (London, 1818).

[2] In *The Philosophy of the Inductive Sciences* (London, 1837). For Whewell as polymath and nomenclator, refer to E. W. Strong, *Journal of the History of Ideas* 16, 209, 1955; P. J. Wexler, *Notes and Queries*, n.s., 8, 27, 1961; S. Ross, *Notes and Records of the Royal Society* 16, 187, 1961. Among the familiar words he invented are *anode, cathode, ion, anion, cation, eocene, miocene, pliocene*, and of course *physicist* and *scientist*. Earlier variants of the latter cited in the *Oxford Dictionary of English Etymology* (ed. C. T. Onions) are *sciencer, scientiate, sciencist* and *scientman*.

their words, fix your attention on their deeds.'[1] Darwin's case is notorious. In his autobiographical sketch,[2] contemporary with the sixth edition of *The Origin of Species*, he said of himself that he 'worked on true Baconian principles, and without any theory collected facts on a wholesale scale'; but later in the same work he said that he could not resist forming a hypothesis on every subject, and he gave away his true opinions (as opposed to the opinions which he felt became him) in letters to Henry Fawcett and H. W. Bates.

Darwin's self-deception is one that nearly all scientists practise, for they are not in the habit of thinking about matters of methodological policy. Ask a scientist what he conceives the scientific method to be, and he will adopt an expression that is at once solemn and shifty-eyed: solemn, because he feels he ought to declare an opinion; shifty-eyed, because he is wondering how to conceal the fact that he has no opinion to declare. If taunted he would probably mumble something about 'Induction' and 'Establishing the Laws of Nature', but if anyone working in a laboratory professed to be trying to establish Laws of Nature by induction we should begin to think he was overdue for leave.

You must admit that this adds up to an extraordinary state of affairs. Science, broadly considered, is incomparably the most successful enterprise human beings have ever engaged upon; yet the methodology that has presumably made it so, when propounded by learned laymen, is not attended to by scientists, and when propounded by scientists is a misrepresentation of what they do. Only a minority of scientists have received instruction in scientific methodology, and those that have done so seem no better off.

One way out of this dilemma is to argue that scientific methodology

[1] 'On the Method of Theoretical Physics', in *The World as I See It* (London, 1935).

[2] In *The Life and Letters of Charles Darwin*, ed. F. Darwin (London, 1887), pp. 83, 103. The letters to Fawcett and to Bates are in *More Letters of Charles Darwin*, ed. F. Darwin and A. C. Seward (London, 1903), pp. 176, 195. To Fawcett he wrote (18 September 1861):

'About thirty years ago there was much talk that geologists ought only to observe and not theorise; and I well remember someone saying that at this rate a man might as well go into a gravel-pit and count the pebbles and describe the colours. How odd it is that anyone should not see that all observation must be for or against some view if it is to be of any service.'

To Bates (22 November 1860):

'I have an old belief that a good observer really means a good theorist.'

is understood intuitively by scientists and needs to be propounded only for the benefit of other people. Nearly all scientists are loud in deploring the utterly unscientific way in which everyone else carries on – politicians, educationalists, administrators, sociologists – and it is upon *them* that they urge the adoption of the scientific method, whatever it may be. John Stuart Mill, the most influential of all methodologists, was certainly not trying to teach scientists their business. On the contrary, his ambition was to analyse and expound their methods in the hope that the complex and baffling problems of society would eventually give way before their use. In a sense this was Bacon's ambition too, for though one cannot be confident of any simplified interpretation of that brilliant and strangely compounded character, yet his *New Atlantis* is the very consummation of what he thought the application of his methods might achieve.

Perhaps, then, we should no longer think of scientific methodology as a discipline of which the chief purpose is to teach scientists how to conduct their business, but rather as an attempt to get non-scientists to pull themselves together and smarten up and in general be much more scientific than they are. Many modern methodological texts have therefore a strong orientation towards the social and behavioural sciences, as if sociologists and social anthropologists were backward because (poor things) they had not been properly brought up in the manners and usages of polite science. While I respect this evangelistic mission, I am not in sympathy with it. The 'backwardness' of sociology (as of biology in the nineteenth century) has little now to do with a failure to use authenticated methods of scientific research in trying to solve its manifold problems. It is due above all else to the sheer complexity of those problems. I very much doubt whether a methodology based on the intellectual practices of physicists and biologists (supposing that methodology to be sound) would be of any great use to sociologists. On the contrary, the influence of inductivism, the subject of Part II, has in the main been mischievous. It has stirred up in some sociologists the ambition to ascertain the laws of social change, above all by the painstaking accumulation of data out of which general principles will in due course take shape. The elevated prose and studied postures of a flourishing school of social anthropology in France today are best explained away as a reaction against the crude scientism of those who have urged upon sociologists the adoption of a style of investigation which they do not use themselves and cannot authenticate from their own experience.

3

I have said so much that is critical of scientific methodology that it may seem strange that I should have chosen to write about it at all. I seem to do nothing but find fault.

If I have given that impression I must at once correct it. Even if it were never possible to formulate *the* scientific method, perhaps because there is no such thing, yet scientific methodology, as a discipline, would still have a number of distinctive and important functions to perform. For in the practice and interpretation of science a number of real problems arise which are common to all sciences but are 'formal' in the sense that they do not depend on what the particular sciences are about. These are ample agenda for a school of methodology: let me now briefly mention three.

1 The problem of *validation*: of the grounds upon which general statements may be judged true or false or merely probable, and of the methods by which we may quantify their degree of imprecision. Under this heading I classify the illuminating developments of modern statistical analysis, particularly in the domain of small-sample theory, so much of it the work of mathematicians turned scientist or of mathematically minded biologists.[1] Matters of validation are important in the experimental sciences, but not as important as they are sometimes made out to be. (I argue in Part II that an obsessional preoccupation with matters to do with ascertainment is part of the heritage of the inductivism.) It is in the *generation* of scientific knowledge, not in its interpretation or in a retrospective analysis of 'the data', that scientists are oppressed by the fear of error. It is a truism to say that a 'good' experiment is precisely that which spares us the exertion of thinking: the better it is, the less we have to worry about its interpretation, about what it 'really' means.

2 *Reducibility; emergence.* If we choose to see a hierarchical structure

[1] e.g. R. A. Fisher, F. W. Yates, 'Student', J. H. Gaddum.

in Nature – if societies are composed of individuals, individuals of cells, and cells in their turn of molecules – then it makes sense to ask whether we may not 'interpret' sociology in terms of the biology of individuals or 'reduce' biology to physics and chemistry. This is a living methodological problem, but it does not seem to have been satisfactorily resolved. At first sight the ambition embodied in the idea of *reducibility* seems hopeless of achievement. Each tier of the natural hierarchy makes use of notions peculiar to itself. The ideas of democracy, credit, crime or political constitution are no part of biology, nor shall we expect to find in physics the concepts of memory, infection, sexuality or fear. No sensible usage can bring the foreign exchange deficit into the biology syllabus, already grievously overcrowded, or nest-building into the syllabus of physics. In each plane or tier of the hierarchy new notions or ideas seem to emerge that are inexplicable in the language or with the conceptual resources of the tier below. But if in fact we cannot 'interpret' sociology in terms of biology or biology in terms of physics, how is it that so many of the triumphs of modern science seem to be founded upon a repudiation of the doctrine of irreducibility? There is a problem here to which methodologists can and do make valuable and illuminating contributions.[1]

3 *Causality*: the problems raised by the notion of necessary connection, and the discussion of its actual and proper use. No one who has studied the slovenly and sometimes actively misleading ways in which geneticists were at one time wont to discuss the relationship between 'gene' and 'character' will dismiss the problem as dead or unworthy of attention. To bring the point home, let me make four successive statements about the role of the Y chromosome in the determination of sex in man. In man

(*a*) the possession of a Y chromosome is the cause of maleness;

[1] See, for example, E. Nagel, *The Structure of Science* (New York, 1961); A. Pap, *An Introduction to the Philosophy of Science* (London, 1963). The problem of 'reducing' sociology to biology goes back at least to John Stuart Mill: 'The laws of the phenomena of society are, and can be, nothing but the laws of the actions and passions of human beings united together in the social state . . . Human Beings in society have no properties but those which are derived from, and may be resolved into, the laws of the nature of individual men.' *A System of Logic* (7th ed., London, 1868), book VI, chapter VII, § 1. For further discussion see Appendix, pp. 111–14 below.

(*b*) the possession of a Y chromosome causes the difference between male and female characteristics;

(*c*) the substitution of a Y chromosome for one of the two X chromosomes causes the difference between male and female characteristics;

(*d*) there is a wide but definable class of genetic and environmental situations in which the substitution of a Y chromosome for one of the two X chromosomes causes the difference between male and female characteristics.

These statements mark four stages in the refinement of a vague and barely articulate but obviously 'significant' idea. The first is scientifically semi-literate; the fourth, though long and clumsy, is pretty well acceptable. The notion of causality pervades the whole of science, and no one science has any special claim to adjudicate upon its usage.[1] The existence of problems of this kind is justification enough for the existence of a science or area of discourse known as scientific methodology, even if its task falls short of expounding the nature of scientific method as a whole. I must not therefore be thought to imply that the pursuit of methodology is a waste of time.

I have entitled this Part 'The Problem Stated', and the problem is twofold: that which is embodied in the question 'What is the scientific method?' and that which is embodied in the fact that scientists pay no serious attention to the answer. But answers have been given, in spite of the scientists' indifference, and in Part II I shall ask whether the doctrine of *induction* provides a good enough approximation to the truth.

[1] The terminology by which we speak of a gene substitution's causing a character difference was quite largely influenced by L. T. Hogben, *Nature and Nurture* (London, 1933). Many think it clumsy and advocate more elaborate formulations (e.g. J. H. Woodger, *Biology and Language* (Cambridge, 1952)). However, the usage is one that comes naturally to experimental scientists. When we carry out an experiment of ordinary unifactorial design (one factor or circumstance varied, the others kept constant), the result of the experiment is the *difference* between two sets of readings (or two sets of phenomena or two events), namely those recorded in the experiment itself and those recorded in its 'control'; and the inference we are entitled to draw is that the difference between the starting conditions was the cause of the difference between the two sets of results. This is precisely the genetic usage. In everyday life, of course, we speak of the causes of events, phenomena or states of affairs, but the cause we have in mind, when analysed, usually turns out to be the cause of a difference between *what was* and *what might have been*; between what did happen and what might have happened if the antecedents had themselves been different.

TOWARDS the end of Part I, I said that scientific methodology had important but limited tasks to perform in the analysis or clarification of certain ideas that were common to all the sciences. I mentioned three of them: the ideas of validation or justification, of reducibility and emergence, and of causality. The older methodologists would not have been satisfied with such limited ambitions. Their intention was to lay bare the whole structure of scientific reasoning, to expound all the distinctive acts of thought that enter into scientific discovery and the enlargement of the understanding.

For more than a hundred years the English-speaking world has been dominated by the opinion that scientific reasoning is of a special kind, *inductive*: an opinion so strongly advocated by such skilful and persuasive thinkers that even when many of the principles of induction have been repudiated or allowed to fade away we still remain in an unconscious bondage to a number of inductive practices and habits of thought.

To the question 'What is induction?' there is no simple answer, even when we eliminate mathematical induction (a special usage) and that humble form of induction which assures us that what is true of each must be true of all. Inductivism is a formulary of beliefs, a complex of attitudes and practices having to do with the nature of science and of scientific enquiry, and in what follows I shall do my best to give a fair account of the various elements that enter into it. Although much of what I shall say will be critical, I do not want to give the impression that, in my opinion, induction has no place in science at all. There are indeed certain limited and special occasions on which we carry out induction according to the rules of Bacon and of Mill, but I shall defer mention of them until later on.

One point should be made clear from the beginning. In the traditional view of induction, that which is embodied in dictionary definitions, induction is 'arguing from the particular to the general' where deduction is arguing from the general to the particular. Induction, then, is a scheme or formulary of reasoning which somehow empowers us to pass from statements expressing particular 'facts' to general statements which comprehend them. These general statements (or laws or principles) must do more than merely summarise the information contained in the simple and particular statements out of which they were compounded: they must add something, say more than that which has been said already – for what would be the use of a 'Law of Nature' which merely authenticated or conferred respectability upon the phenomena already known to obey it? Inductive reasoning is *ampliative* in nature. It expands our knowledge, or at all events our pretensions to knowledge.

This is all very well, but the point to be made clear is that induction, so conceived, cannot be a logically rigorous process. It cannot (as deduction can, if properly executed) lead us with certainty to the truth. Mill believed it could do so, but John Venn and C. S. Peirce and others flatly disagreed with him,[1] and it is their opinion that has prevailed. I shall waste no time attacking a position that is no longer defended. No process of reasoning whatsoever can, with logical certainty, enlarge the empirical content of the statements out of which it issues. If it could indeed do so then all scientific research could be carried out in a recumbent posture, with the eyes half closed.

2

Now let me discuss one by one the shortcomings of the inductive style of reasoning, as I see it; for in finding fault with induction I am by implication helping to define the properties which a really adequate methodology should enjoy.

§ 1   At the very heart of induction lies this innocent-sounding belief: that the thought which leads to scientific discovery or to the propounding of a new scientific theory is logically accountable and can be logically spelled out. Even if they are not apparent at the time (because they have been short-circuited or speeded up), a retrospective analysis can reveal the processes of reasoning and the logically motivated actions that conduct the scientist towards what he believes to be the

---

[1] Amongst the others were Dugald Stewart (writing fifty years before Mill) and Stanley Jevons. On this point, see pp. 119–21 below.

truth. There is a grammar of science, and the language of science can
be parsed.

This is quite different from saying that *given* some belief or opinion
or would-be natural law, no matter what its origin (whether by research, by revelation, or in a dream), *then* its acceptability can be tested
by procedures that involve the use of logic. In the inductive view, it is
the process of *getting* an idea or formulating a general proposition that
can be logically reasoned out. It follows that, in the inductive scheme,
discovery and justification form an integral act of thought. That
which leads us to form an opinion is also that which justifies our holding the opinion. The intellectual processes that conduct us towards a
generalisation are themselves the grounds for supposing it to be true.[1]

This concept of the inductive process must have arisen out of a misleading formal analogy with *d*eduction. In deductive reasoning, for
example in Euclid, we discover or uncover a theorem by reasoning
which, if we have carried it out correctly, guarantees the theorem to
be true if the axioms or premises are true. Our ability to deduce
Pythagoras' Theorem from Euclid's axioms – that is, to discover
Pythagoras' Theorem in Euclid's axioms – is in itself our justification
for believing it to be valid. In a purely formal sense, therefore, discovery and justification are the same process in deductive logic; but the
qualification 'in a purely formal sense' is very important. It is most
unlikely that more than a tiny minority of mathematical theorems were
ever in fact arrived at, 'discovered', merely by the exercise of deductive
reasoning. Most of them entered the mind by processes of the kind
vaguely called 'intuitive'; deduction or logical derivation came later,
to justify or falsify what was in the first place an 'inspiration' or an
intuitive belief. This is seldom apparent from mathematical writings,

[1] See Whewell's critique of Mill in chapter 22, particularly pp. 262ff., of
The Philosophy of Discovery (London, 1860). The later editions of Mill's *System*
challenge Whewell's objections (book III, chapter VI, § 6), though not very con-
vincingly. Mill ends rather lamely by saying, with reference to his Four Methods:
'If discoveries are ever made by observation and experiment without Deduction, the four methods are methods of discovery: but even if they were not
methods of discovery, it would be not less true that they are the sole methods of
Proof . . .'.

I feel that Mill was handicapped by his deeply held conviction that induction
and deduction were processes of the same stature or logical authority, and in the
next paragraph of the main text I try to show how misleading this belief can be.

because mathematicians take pains to ensure that it should not be. Deductivism in mathematical literature and inductivism in scientific papers are simply the postures we choose to be seen in when the curtain goes up and the public sees us. The theatrical illusion is shattered if we ask what goes on behind the scenes. In real life discovery and justification are almost always different processes, and a sound methodology must make it clear that they are so.

§ 2   Inductive theory insists on the primacy of Facts: of propositions that put on record the simple and uncomplicated evidence of the senses. Karl Pearson[1] was a great believer in facts:

The classification of facts and the formation of absolute judgments upon the basis of this classification . . . essentially sum up the *aim and method of modern science*.

The classification of facts, the recognition of their sequence and relative significance, is the function of science.

Pearson felt that the study of facts was conducive not only to good science but to right-mindedness in general:

Modern Science, as training the mind to an exact and impartial analysis of facts, is an education specially fitted to promote sound citizenship.

It may therefore seem downright subversive to question the primitive authenticity of facts or to cast doubt upon evidence in the form in which it is delivered to us by the senses; it is worse still to ask, as Whewell did, how often 'facts' can be stripped of a mask of interpretation and theory. It is very un-English, to be sure, for to put such a question is to challenge the greatest philosophic tradition of the English-speaking world, the tradition of philosophic empiricism which we inherit from John Locke. Nothing enters the mind except by way of the senses (its fundamental principle goes); and though the senses may sometimes be clouded, though we may sometimes be the victims of deception and illusion, yet if we can only get at it in its primitive simplicity, the evidence of the senses is the foundation of all knowledge that is not merely a repartition of ideas or words. There is an *essential*

---

[1] *The Grammar of Science* (3rd ed., London, 1911; 1st ed., 1892). The passages I quote are from pp. 6 and 9 (see also p. 37). The italics are Pearson's.

trustworthiness about the evidence of the senses, and therefore about the simple observational statements which put that evidence on record.

It won't do, of course. No one now seriously believes that the mind is a clean slate upon which the senses inscribe their record of the world around us: that we take delivery of the evidence of the senses as we take delivery of the post. 'Everything that reaches consciousness is utterly and completely adjusted, simplified, schematised, interpreted,' said Nietzsche,[1] in one of his exhilarating outbursts of common sense. Innocent, unbiased observation is a myth: 'experience is itself a species of knowledge which involves understanding,' said Kant.[2] What we take to be the evidence of the senses must itself be the subject of critical scrutiny. Even the fundamental principle of empiricism is open to question, for not all knowledge can be traced back to an origin in the senses. We inherit some kinds of information. A bird's song is in some sense the transcription of a chromosomal tape recording, and the same goes for the entire repertoire of all that can properly be called 'instinctual' behaviour.[3]

[1] Friedrich Nietzsche, *The Will to Power*, trans. A. M. Ludovici (London, 1910). See especially vol. 2, §§ 466–617. Nietzsche's ideas on methodology are worth serious attention; those who have dismissed him as the author of *Thus Spake Zarathustra* will be surprised at the strength and insight of his opinions on methodology and the theory of knowledge.

[2] *The Critique of Pure Reason*, trans. N. Kemp Smith (2nd ed., London, 1933), p. 22.

[3] To my mind the most striking examples of 'genetically programmed' behaviour-sequences are not those in which an animal reared in isolation from birth or hatching turns out to 'know' how to sing a particular song or build a nest, but those in which a suitable stimulus transforms one behaviour pattern into another. A wealth of examples relating to sexual behaviour will be found in D. S. Lehrman, *Sex and Internal Secretion* (3rd ed., Baltimore, 1961), pp. 1268–382. Thus (p. 1299):

'Domestic cocks take no part in the care of the young, and sometimes even kill chicks that are confined with them. [Cocks cannot be induced to incubate eggs by prolactin injections, but] a number of workers have reported that prolactin induces cocks to cluck, to lead, and to protect chicks under their body and wings.'

As to the level of detail of the genetic programme, we may note (p. 1351) that

'Courting male canaries sometimes dangle a piece of string or cotton before the female. Shoemaker . . . found that female canaries injected with testosterone propionate postured like courting males, and also engaged in this string carrying behaviour, reminiscent of the carrying of nesting material.'

The argument that all instinctual behaviour must at some time have been

§ 3   Although inductive exercises often begin with an injunction to assemble all the 'relevant' information (relevant to what?), inductive theory provides no *formal* incentive for making one observation rather than another. Why indeed do we not count and classify the pebbles in a gravel pit (see p. 80, note 2)? This is a subject on which Coleridge had a number of pointed and sensible things to say.[1] Any adequate account of scientific method must include a theory of incentive or special motive; must contain a canon to restrict observation to something less than the whole universe of observables. We cannot browse over the field of nature like cows at pasture.

§ 4   I said in Part I (with induction in mind) that scientific methodology was almost obsessively preoccupied with problems of justification and ascertainment – with laying down the conditions under which the views we hold should be judged right or wrong, or with quantifying the degree of confidence we should have in them. When scientific research is studied on the hoof, so to speak, we find that very few theories are utterly discredited in the style in which (for example) Thomas Henry Huxley demolished Goethe's and Oken's Vertebral Theory of the skull.[2] Theories are repaired more often than they are refuted, and a methodology of rectification (a logical variant of negative feedback) is something we shall expect to find in any satisfactory formal account of scientific reasoning. Sometimes theories merely fade away: no one now believes in the doctrine of 'Protoplasm', but no one to my knowledge has ever refuted it.[3] More often they are merely assimilated into wider theories in which they rank as special cases. The Law of Recapitulation and the Germ-Layer Theory have not been shown to be 'wrong'. They have simply lost their identity and their

---

learned, though not necessarily in the generation in which it is acted out, will not hold water, for a chromosomal aberration might conceivably produce a purely endogenous change of behaviour pattern. For an early discussion of 'inherited ideas', see chapter IV in E. Mach, *The Analysis of Sensations*, trans. C. M. Williams and S. Waterlow (London, 1914; 1st ed., 1885).

[1] Coleridge *On Method* (see p. 79 above, note 1). See also p. 126 below.
[2] In his famous Croonian Lecture to the Royal Society: *Proceedings of the Royal Society*, series B, 9, 381–457, 1857–9 (reprinted in *The Scientific Memoirs of T. H. Huxley*, ed. M. Foster and E. Ray Lankester (London, 1898), vol. 1, pp. 538–606).
[3] See p. 293 below.

special significance in an improved understanding of the mechanism of development. They have been trivialised.[1]

§ 5 One of the most damaging charges against inductivism was brought for the first time, I believe, by Lord Macaulay,[2] though not in the form in which I shall present it here. Inductivism gives no adequate account of scientific fallibility, fails altogether to explain how it comes about that the very same processes of thought which lead us towards the truth lead us so very much more often into error.

Methodologists who have no personal experience of scientific research have been gravely handicapped by their failure to realise that nearly all scientific research leads nowhere – or, if it does lead somewhere, then not in the direction it started off with. In retrospect we tend to forget the errors, so that 'The Scientific Method' appears very

[1] *Recapitulation*. If a human being has embryonic gill pouches before his lungs begin to form, that is not because development is obliged to rehearse evolution, but rather because the lung *is* a special sort of gill derivative, just as jaws are gill arches of a special kind, or limbs a special sort of fin. It is therefore understandable from the point of view of developmental mechanics that the embryos of higher vertebrates should sometimes recapitulate the *embryonic* history of their ancestors (see G. R. de Beer, *Embryos and Ancestors* (Oxford, 1940)). The Law of Recapitulation is to some extent true of what might be called 'serial' evolution, the substitution of one terminal developmental episode for another, as in the examples cited above; but there is no reason why it should be expected to be true of evolutionary processes that foreshorten the life history (neoteny) or which introduce novelties during embryonic or juvenile stages of development.

Germ-Layer Theory. The embryos of many vertebrates have a layered or laminate structure, and there is a surprising degree of regularity in the nature of the adult organs into which the several layers develop – outermost layer into skin and nervous system, innermost layer into the viscera, and so on. The layered structure, as we see it under the microscope, now merely reminds us of something unknown to the embryologists who propounded the germ layer theory, viz. that the morphogenetic exercises of early vertebrate development consist of the sliding, folding, stretching and glassblower-like manipulations of cellular sheets as they come to take up the remarkably uniform stations characteristic of the early axiate embryo of vertebrates.

The Law of Recapitulation and the Germ-Layer Theory are not 'wrong': they can be used to make a point about development, even if the point is now hardly worth making. We call them to mind only to help us understand the newer reasoning by which they have been superseded, for something akin to a principle of recapitulation works here too, in the history of ideas.

[2] In his extended review (often reprinted in collections of Macaulay's 'Essays') of Montagu's edition of Bacon's *Works* in the *Edinburgh Review*, July 1837.

much more powerful than it really is, particularly when it is presented to the public in the terminology of breakthroughs, and to fellow scientists with the studied hypocrisy expected of a contribution to a learned journal. I reckon that for all the use it has been to science about four-fifths of my time has been wasted, and I believe this to be the common lot of people who are not merely playing follow-my-leader in research.

Why do scientists hold or come to formulate erroneous opinions? That, surely, is a central problem of methodology. For traditional inductive theory, as Popper has explained,[1] error must be held to arise from misapprehension of the facts, the data; from a misreading of that Book of Nature in which the truth resides if only we had the skill and clarity of vision to discern it. If the facts are misapprehended through blindness or prejudice, then the inferences logically induced from them must be mistaken; and so we are led into error. (This follows the deductive analogy pretty closely, for if in deduction the theorems are empirically wrong it must be because the axioms were empirically wrong, unless there has been an avoidable mistake of reasoning.) If only we had a clear and perspicacious vision the truth would make itself apparent to us, as it was to the chap in Voltaire who apprehended all human knowledge in a matter of weeks because, until he grew out of savage innocence, his mind had been preserved from the prejudices and ideological corruptions of prior learning.[2]

This cannot be the whole truth about error. What shows a theory to be inadequate or mistaken is not, as a rule, the discovery of a mistake in the information that led us to propound it; more often it is the contradictory evidence of a new observation which we were led to make *because* we held that theory. Error or insufficiency is shown up by a critical process applied in retrospect; only seldom is it due to a failure of apprehension, to a dullness or cloudiness of vision. It is a bad mark

[1] Karl Popper, *Conjectures and Refutations* (London, 1963), pp. 3–30.

[2] I came across Voltaire's *L'Ingénu* through J. Agassiz's sparkling essay 'Towards an Historiography of Science' (*History and Theory*, 2, 1963). Brought up in feral innocence, the hero 'faisait des progrès rapides dans les sciences . . . La cause du développement rapide de son esprit était due à son éducation sauvage presque autant qu'à la trempe de son âme: car, n'ayant rien appris dans son enfance, il n'avait point appris de préjugés. Son entendement, n'ayant été courbé par l'erreur, était demeuré dans toute sa rectitude. Il voyait les choses comme elles sont . . .' (chapter XIV). *L'Ingénu* was Inductive Man.

against inductivism that it provides us with no acceptable theory of the origin and prevalence of error.

§ 6   What are we to make of *luck* in our methodology of science? In the inductive view, luck strikes me as completely inexplicable; it can arise only from the gratuitous obtrusion of something utterly unexpected upon the senses; it is like winning a prize in a lottery in which we did not buy a ticket. To buy a ticket is to define a category of expectations, and then the reason why we won is obvious: we were in luck; for once in a way our hopes were gratified. We have Fontenelle's and Pasteur's word for it that luck makes sense only against a background of prior expectations. Ever since his experiences in the First World War, Alexander Fleming had been deeply concerned by the problem of infected wounds. It was his lifelong ambition to discover a non-toxic antibacterial agent, and in penicillin he found a winner – by luck, if you like; but he held a ticket which entitled him to win a prize.[1]

§ 7   In § 5 I said that the shortcomings of scientific theories were usually revealed by the exercise of a critical process. Classical inductive theory reveals no clear grasp of the *critical* function of experimentation. 'Experiments' are of several different kinds – in a moment I shall mention four – and what I shall describe as inductive or 'Baconian' experimentation is only one of them. Let me first explain what I conceive Baconian experimentation to be.

If we believe that the initiative for scientific research lies in observation, that scientific knowledge grows out of the evidence of the senses, then our first duty as scientists must be to observe nature faithfully, intently, and without misleading preconceptions. Let us then imagine ourselves wandering through a sylvan or pastoral world and recording our observations. Obviously we could not observe enough to sustain the growth of science. We could spend a lifetime observing nature without once seeing two sticks rub together, seeing sunlight refracted through a crystal, or witnessing the distillation of fermented liquor. Francis Bacon therefore charged us to contrive or invent experiences; to mess about, we might say today (the phrase would not have come

[1] There was indeed an element of blind luck in the discovery of penicillin. See my review of J. D. Watson's *The Double Helix*, pp. 270–8 below.

naturally to Bacon); to indulge in experiential play. Let us try experiments[1] with the burning glass and loadstone and rubbed amber; let us distil liquors not once but twice successively, to see what happens. Nature is an actress with a prodigious repertoire: give her an opportunity to perform.

It was these contrived experiences, invented happenings, that Bacon called *experiments*, and this is still the vernacular usage of today.[2] Experimentation in Bacon's sense is not a critical procedure. Its purpose is to nourish the senses, to enrich the repertoire of factual information out of which inductions are to be compounded. It is that enlargement of experience which, in the inductive view, cannot but lead to an enlargement of the understanding.

Experiments are of at least four kinds:

(i) Inductive or Baconian experiments such as those I have just described ('I wonder what would happen if . . .').

(ii) Deductive or Kantian experiments, in which we examine the consequences of varying the axioms or presuppositions of a scheme of deductive reasoning ('Let's see what happens if we take a different view'). In introducing the greatest intellectual exploit in the history of philosophy, Kant invites us, 'by way of an experiment in pure reason', to abandon the commonplace view that our sensory intuitions conform to the constitution of objects, and to examine the consequences of supposing that objects conform to the constitution of our faculty of intuition. 'This experiment succeeds as well as could be desired'; it explains how we may have knowledge that is independent of all experience, and leads Kant to propose that space and time are not objects of but *forms* of sensory intuition.[3]

Perhaps the best examples of 'pure' Kantian experiments are those

---

[1] Notice the terminology: nowadays we *do* or *carry out* experiments; we no longer *make* or *try* experiments.

[2] Mill's usage is essentially the same, though it is naturally more sophisticated: see his *System*, book III, chapter VII, §§ 3, 4. 'Artificial' experimentation is an extension of 'pure' observation: 'it enables us to obtain innumerable combinations of circumstances which are not to be found in nature, and so add to nature's experiments a multitude of experiments of our own'.

[3] Kant, op. cit. (p. 89 above, note 2). To call such experiments 'Kantian' is by no means to imply that Kant did not have a clear understanding of the critical functions of experimentation; Kant's grasp of scientific methodology was remarkable (see also p. 103 below, note 1).

which led to the formulation of the classical non-Euclidian geometries (hyperbolic, elliptic) by Saccheri, Lobachevsky, Bolyai, Riemann and others.[1] It is ironical that this should be so, for as W. K. Clifford cleverly discerned,[2] the non-Euclidian geometries might be thought to call into question the very truths concerning a priori knowledge to which Kant supposed that his own audacious experiment had led him.

(iii) Critical or Galilean experiments: actions carried out to test a hypothesis or preconceived opinion by examining the logical consequences of holding it. Galilean experiments discriminate between possibilities. I shall say more about them in Part III.

(iv) Demonstrative or Aristotelian experiments, intended to illustrate a preconceived truth and convince people of its validity. In *Plus ultra* (1668), a puff for the New Science, Joseph Glanvill wrote of Aristotle that 'he did not use and imploy Experiments for the erecting of his Theories: but having arbitrarily pitch'd his Theories, his manner was to force Experience to suffragate, and yield countenance to his precarious Propositions'.[3] Thomas Sprat took a very poor view of experimentation in this style: 'a most venomous thing in the making of sciences; for whoever has fix'd on his Cause, before he has experimented, can hardly avoid fitting his Experiment to his own Cause . . . rather than the Cause to the truth of the Experiment it self'.[4]

---

[1] We may construe the classical non-Euclidian geometries as being derivable by the experiment of replacing Euclid's axiom of parallels (or its formal equivalents) by the 'obtuse angle hypothesis' or the 'acute angle hypothesis' (or their formal equivalents), so generating geometries representable in spaces of constant positive curvature or constant negative curvature respectively (see W. K. Clifford, note 2 below). In this scheme Euclidian geometry is a special case, representable in a space of zero curvature. It has often been pointed out that the formulation of non-Euclidian geometries had a profound influence on our conception of mathematical reasoning and the nature of the 'truths' to which it leads.

[2] See in particular W. K. Clifford's four lectures on 'The Philosophy of the Pure Sciences' (*Lectures and Essays*, ed. Leslie Stephen and F. Pollock (London, 1879), vol. I, pp. 254–340); see also his lecture 'On the Aims and Instruments of Scientific Thought', ibid., pp. 124–57. Clifford had a deep understanding of mathematical thought and his lectures are a delight to read, but his views on the empirical sciences are not always convincing. His criticism of Kant must be reappraised in the light of our present knowledge that Kant was 'in on' the earliest stages of speculation about the necessary validity of Euclidian geometry: see A. C. Ewing, *A Short Commentary on Kant's Critique of Pure Reason* (London, 1938).

[3] p. 112.   [4] *History of the Royal Society* (1667), p. 108.

Most original research begins with Baconian experimentation. We undertake to study a problem or a phenomenon for many different reasons: because it is interesting or important, because we have been led to it through earlier researches, or because we have been asked or told to do so; but whatever our motives may have been, the first thing to do is to find out what actually happens, what is in fact the case ('Let's find out in a bit more detail what it is we are actually studying'). We describe and annotate the phenomena when they are made to take place under certain well-defined and well-regulated conditions. In the meanwhile we begin to form opinions about the nature of the causal mechanisms at work and the relationship of the phenomena to others, and only critical experimentation can discriminate between them. Sciences which remain at a Baconian level of development, as (for example) the study of animal behaviour did until its modern renaissance,[1] amount to little more than academic play.

I said near the beginning of this Part that there were certain situations in which we undoubtedly use inductive reasoning – and in very much the style envisaged by Bacon or, more explicitly, by Mill. Suppose we have set up some 'tissue cultures' of living cells, using a variety of media which have subsequently been thrown away. Some of the cultures, but not all, have been ruined by bacterial infection, and we naturally wish to find out why. Mill's Five Canons[2] will conduct us towards the answer; or, if they do not, nothing will. Media common to all the cultures cannot have been responsible for introducing the infection. If the infected cultures, and they alone, were set up with a medium from a certain special source, then that medium was almost certainly responsible; and we shall be confirmed in this interpretation if we find that the more heavily contaminated cultures were those in which a larger quantity of the medium under suspicion had been used. We are taken aback when a fuller study of the records shows that a number of cultures escaped infection although the supposedly infected medium had been used to prepare them, but it turns out that these anomalous cultures differed from those which were overtly contaminated by the use of a bactericidal ingredient which kept the infection down. And so on: the situation can be made as complicated as we please, but the reasoning which resolves it is straightforward and quite commonplace.

[1] See p. 295–7 below.   [2] *System*, book III, chapters VIII and IX.

It is 'logical' in the sense that it can be carried out by formula or by rote, and it is, or can be, conclusive if the empirical facts, as stated, represent the whole truth and nothing but the truth. Conversely, if our conclusions are wrong it must be because there was a mistake in the 'facts' from which the induction started.[1]

We turn naturally to inductive reasoning when we undertake a retrospective causal analysis of a state of affairs that is already 'given'. Was it not perhaps the goal of inductivism that all scientific reasoning should aspire to such a condition? Let us first assemble the data; let us by observation and by making experiments compile the true record of the state of Nature, taking care that our vision is not corrupted by pre-conceived ideas; *then* inductive reasoning can go to work and reveal laws and principles and necessary connections. A travesty of inductivism, you may think; but with something taken away for rhetorical overstatement, I think it not far short of the truth.

Let me now bring together the main points I have tried to make in this second Part. Inductivism is a complex of opinions about the origin and character of scientific reasoning, and though it all hangs together in a creaking and groaning way it also has a number of ruinous

---

[1] Mill's own examples of induction at work are strange reading today (ibid., note 30), particularly when he chooses biological examples, as Whewell pointed out (*Philosophy of Discovery*, chapter XXII, § 41). Mill's methods, he adds, have a great resemblance to Bacon's, and we may note that Macaulay's scathing example of inductive reasoning (p. 91 above, note 2) antedates both Whewell and Mill. It is no different in principle from my infected tissue cultures, but here the infection is gastrointestinal (if indeed an allergy is not to blame):

'A plain man finds his stomach out of order. He never heard Lord Bacon's name. But he proceeds in the strictest conformity with the rules laid down in the second book of the *Novum organum*, and satisfies himself that minced pies have done the mischief. "I ate minced pies on Monday and Wednesday, and I was kept awake by indigestion all night." This is the *comparentia ad intellectum instantiarum convenientum*. "I did not eat any on Tuesday and Friday, and I was quite well." This is the *comparentia instantiarum in proximo quae natura data privantur*. "I ate very sparingly of them on Sunday, and was very slightly indisposed in the evening. But on Christmas Day I almost dined on them, and was so ill that I was in great danger." This is the *comparentia instantiarum secundum magis et minus*. "It cannot have been the brandy which I took with them. For I have drunk brandy daily for years without being the worse for it." This is the *rejectio naturarum*. Our invalid then proceeds to what is termed by Bacon the *Vindemiatio*, and pronounces that minced pies do not agree with him.'

Here too is a retrospective causal analysis of 'data', i.e. of information given. For 'minced pies' read 'mince pies' throughout.

shortcomings, the study of which (I said) should help us to plan the prospectus of a sounder methodology. In particular:

(1) Inductivism confuses, and a sound methodology must distinguish the processes of discovery and of justification.

(2) The evidence of the senses does not enjoy a necessary or primitive authenticity. The idea, central to inductive theory, that scientific knowledge grows out of simple unbiased statements reporting the evidence of the senses is one that cannot be sustained.

(3) A sound methodology must provide an adequate theory of special incentive – a reason for making one observation rather than another, a restrictive clause that limits observation to something smaller than the universe of observables.

(4) Too much can be made of matters of validation. Scientific research is not a clamour of affirmation and denial. Theories and hypotheses are modified more often than they are discredited. A realistic methodology must be one that allows for repair as readily as for refutation.

(5) A good methodology must, unlike inductivism, provide an adequate theory of the origin and prevalence of error . . .

(6) . . . and it must also make room for luck.

(7) Due weight must be given to experimentation as a critical procedure rather than as a device for generating information; to experimentation as a method of discriminating between possibilities.

Inductivism falls far short of being an adequate scientific methodology, and in Part III I shall study the credentials of an altogether different scheme of thought.

III   MAINLY ABOUT INTUITION

I

I DEVOTED Part II to finding fault with induction, excusing myself as
I did so on the grounds that I was working my way towards the prospectus of a useful and realistic methodology. Of inductivism, I said
that it was a complex formulary of methodological practices and beliefs
of which the two most important elements were these:

(*a*) Observation is the generative act in scientific discovery. For all
its aberrations, the evidence of the senses is essentially to be relied upon
– provided we observe nature as a child does, without prejudices and
preconceptions, but with that clear and candid vision which adults lose
and scientists must strive to regain.

(*b*) Discovery and justification make one act of reasoning, not two.
Inductive logic embodies a rite of discovery and of proof. Scientific
inference can always be made logically explicit, in retrospect if not at
the time it was actually carried out.

So much of what I have said has been abstract that I feel I should now
make amends with a couple of methodological caricatures.

Consider the act of clinical diagnosis. A patient comes to his physician
feeling wretched, and the physician sets out to discover what is wrong.
In the inductive view the physician empties his mind of all prejudices
and preconceptions and *observes* his patient intently. He records the
patient's colour, measures his pulse rate, tests his reflexes and inspects
his tongue (an organ that seldom stands up to public scrutiny). He
then proceeds to other, more sophisticated actions: the patient's urine
will be tested; blood counts and blood cultures will be made; biopsies
of liver and marrow are sent to the pathology department; tubing is
inserted into all apertures and electrodes applied to all exposed surfaces. The factual evidence thus assembled can now be classified and
'processed' according to the canons of induction. A diagnosis (for
example 'It was something he ate') will thereupon be arrived at by
reasoning which, being logical, could in principle be entrusted to a
computer, and the diagnosis will be the right one unless the raw
factual information was either erroneous or incomplete.

Grossly exaggerated? Of course: as I said, a caricature; but, like a caricature, not exaggerated beyond all reason. It is obviously incomplete because no place has been found for flair and insight, and the enrichment that long experience brings to clinical skills. In Commencement Addresses and other uplifting declarations, clinicians who discourse upon the 'spirit of medicine' will always point out that, while there is a large and profoundly important scientific element in the practice of medicine, there is also an indefinable artistry, an imaginative insight, and medicine (they will tell us) is born of a marriage between the two. But then (it seems to me) the speaker spoils everything by getting the bride and groom confused. It is the unbiased observation, the apparatus, the ritual of fact-finding and the inductive mumbo-jumbo that the clinician thinks of as 'scientific', and the other element, intuitive and logically unscripted, which he thinks of as a creative art.

To see whether this apportionment of credit is a just one, let us turn to another clinician in the act of diagnosis. The second clinician always observes his patient with a purpose, with an idea in mind. From the moment the patient enters he sets himself questions, prompted by foreknowledge or by a sensory clue; and these questions direct his thought, guiding him towards new observations which will tell him whether the provisional views he is constantly forming are acceptable or unsound. Is he ill at all? Was it indeed something he ate? An upper respiratory virus is going around: perhaps this is relevant to the case? Has he at last done his liver an irreparable disservice? Here there is a rapid reciprocation between an imaginative and a critical process, between imaginative conjecture and critical evaluation. As it proceeds, a hypothesis will take shape which affords a reasonable basis for treatment or for further examination, though the clinician will not often take it to be conclusive.[1]

This is a travesty too. The imagination cannot work *in vacuo*: there must be something to be imaginative about, a background of observation and Baconian experimentation, before the exploratory dialogue can begin. Nor have I explained the natural *progression* of thought that

---

[1] Someone who goes to a clinic for a 'check-up' will be exposed to a battery of tests and observations and may well think himself the subject of an inductive exercise. What is in fact happening is a sequential evaluation of likely hypotheses about what is or could be wrong with him.

goes into clinical examination. But if I were asked which of the two accounts of the matter I thought the more helpful and realistic, I should without hesitation say the second. It is very far removed from induction, and belongs to an altogether different lineage of thought.[1] We find hints of it in Robert Hooke and Stephen Hales and Robert Boscovich. There are passages in Kant's lectures which reveal a clear understanding of what we have come to call the *hypothetico-deductive* method (I cite one of them below). By the middle of the nineteenth century it had established itself as the official alternative to the induction advocated by Mill. Whewell's *Philosophy of the Inductive Sciences*[2] (that is, of the empirical sciences) is a masterpiece. Stanley Jevons (*The Principles of Science*)[3] can still be read with profit today; so also C. S. Peirce[4] and Claude Bernard.[5] The principal modern advocate and analyst of the theory is Karl Popper;[6] my indebtedness to his writings will be obvious to anyone who knows them, but he must on no account be held responsible for anything in what follows that may seem unsound.

According to this second view, science, in its forward motion, is not *logically* propelled. Scientific reasoning is an exploratory dialogue that can always be resolved into two voices or two episodes of thought, imaginative and critical, which alternate and interact. In the imaginative episode we form an opinion, take a view, make an informed guess, which might explain the phenomena under investigation. The generative act is the formation of a hypothesis: 'we must entertain some hypothesis,' said Peirce,[7] 'or else forgo all further knowledge', for hypothetical reasoning 'is the only kind of argument which starts a

[1] I sketch the history of some of the main elements of the hypothetico-deductive system in 'Hypothesis and Imagination', pp. 115–35 below, and so do not need to go over the same ground here.

[2] (1st ed., London, 1840; 2nd ed., 1847). *The Philosophy of Discovery* (London, 1868) covers some of the same ground, but is not a substitute for the earlier work.

[3] (1st ed., London, 1873; 2nd revised ed., 1877).

[4] *Collected Papers*, ed. C. Hartshorne and P. Weiss (Harvard, 1932–5).

[5] See p. 73 above, note 1.

[6] *The Logic of Scientific Discovery* (London, 1959), a translation (with new appendices and footnotes) of *Logik der Forschung* (Vienna, [1934]); *Conjectures and Refutations* (London, 1963). See also *The Critical Approach in Science and Philosophy*, ed. M. Bunge (London, 1964); M. Bunge, *Scientific Research* (2 vols, New York, 1967).

[7] op. cit. (note 4 above).

new idea'. The process by which we come to formulate a hypothesis is not illogical but non-logical, that is, outside logic. But once we have formed an opinion we can expose it to criticism, usually by experimentation; this episode lies within and makes use of logic, for it is an empirical testing of the logical consequences of our beliefs. 'If our hypothesis is sound,' we say, 'if we have taken the right view, then it follows that . . .' – and we then take steps to find out whether what follows logically is indeed the case. If our predictions are borne out (logical, not temporal predictions) then we are justified in 'extending a certain confidence to the hypothesis' (Peirce again). If not, there must be something wrong, perhaps so wrong as to oblige us to abandon our hypothesis altogether.

A scientific theory can be thought of as a complex, logically-bonded polymer built out of the following elementary or monomeric form:[1]

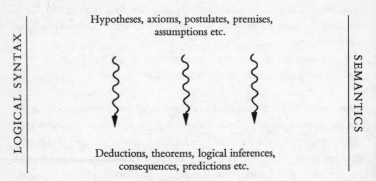

This elementary theory is supported by a *metatheory* which specifies the rules of deduction or statement-transformation ('logical syntax') and adjudicates upon the meanings of the empirical terms which it employs, that is, says what they stand for ('semantics'). Postulates, axioms, premises etc. are statements of the same logical standing; they differ one from another in the ways in which they have come to be formulated and the degree of confidence they enjoy. We assert a postulate and take an axiom for granted, but hypotheses we merely venture to propose. (We *believe* in hypotheses, of course, but only for the sake of argument and as an incentive to critical enquiry: 'belief', as Kant defined it, 'is a kind of consciously imperfect assent'.) The term

[1] See J. H. Woodger, *The Technique of Theory Construction* (Chicago, 1939).

'premise' is coming to acquire an antiquarian flavour. The Revd Sydney Smith, a famous wit, was walking with a friend through the extremely narrow streets of old Edinburgh when they heard a furious altercation between two housewives from high-up windows across the street. 'They can never agree,' said Smith to his companion, 'for they are arguing from different premises.' I explain the point of this story later on. Hypotheses and axioms may be shared between cognate theories, and the logical consequences of one theory may represent the starting-point – the hypotheses or assumptions – of a theory of lower level. It is in this sense that complex theories may be described as logically bonded or logically articulated structures.

Scientific reasoning, in William Whewell's view, is a constant inter-play or interaction between hypotheses and the logical expectations they give rise to: there is a restless to-and-fro motion of thought, the formulation and rectification of hypotheses, until we arrive at a hypothesis which, to the best of our prevailing knowledge, will satis-factorily meet the case.

2

Before discussing the 'hypothetico-deductive' scheme in greater detail, let me call attention to some of its philosophical implications.

If we accept the idea that scientific reasoning is a kind of dialogue between the possible and the actual, between what might be and what is in fact the case, then we are narrowing down the domain of science to one subclass of all possible contentions or beliefs, namely to those which are in principle capable of being modified by critical scrutiny. For, as Kant is reported to have said, it must *certainly* be true of every hypothesis that it could *possibly* be true:[1] hypotheses must be statements of such a kind that they *could* be true. During the 1930s, logical posi-tivists used 'verifiability in principle' as an operational criterion of meaning, one that served to distinguish significant statements from the various allotropic forms of nonsense.

[1] See Kant's *An Introduction to Logic*, trans. C. K. Abbott (London, 1885), pp. 75–6. With Kant's approval, the text was compiled by a student from Kant's lecture notes. Kant's view of the nature and use of hypotheses is very much in the modern style.

It is not now generally believed that 'verifiability' is a property that will serve this purpose. In the first place 'falsifiability' should be substituted for 'verifiability', as Popper has recommended.[1] The arrows in my diagram are intended to signify the polarity or one-wayness of deductive reasoning. Deduction, properly carried out, guarantees that if our hypotheses (axioms, assumptions etc.) are true, then so also, necessarily, will be the inferences drawn from them. If therefore a hypothesis leads to expectations which are not borne out, something *must* be wrong. But if our expectations are fulfilled, it does not by any means follow that the hypotheses which gave rise to them are true, for false hypotheses can lead to true conclusions. We may accept the hypothesis 'as if it were perfectly certain', but 'hypotheses always remain hypotheses, i.e. suppositions to the full certainty of which we can never attain' (Kant).[2]

In the second place (again following Popper), 'falsifiability' marks the distinction between, on the one hand, statements that belong in science and to the world of common sense, and, on the other hand, statements which, though they belong to some other world of discourse, are not to be dismissed contemptuously as nonsense. Metaphysics is a compost that can nourish the growth of scientific ideas. But if we accept falsifiability as a line of demarcation, we obviously cannot accept into science any system of thought (for example, psychoanalysis) which contains a built-in antidote to disbelief: that to discredit psychoanalysis is an aberration of thought which calls for psychoanalytical treatment. (The critic cannot win against such a contention – but he does not have to compete.)

As Whewell pointed out (though Mill, rather perversely I think, disagreed with him),[3] the extramural implications of a hypothesis are often specially valuable – the new expectations it gives rise to, lying outside the phenomena it was originally intended to explain – because they make it possible to expose the hypothesis to independent or at least to unpremeditated tests. 'Predictions' of this kind, when they

---

[1] See p. 101 above, note 6, and p. 127 below.   [2] See p. 103 above, note 1.
[3] See pp. 127, 130 below. For the argument between Mill and Whewell, see Whewell, *The Philosophy of the Inductive Sciences* (2nd ed.), pp. 62–4; Mill, *A System of Logic* (later editions), book III, chapter XIV, § 6; Whewell, *The Philosophy of Discovery*, pp. 272–4.

come off, are a source of special satisfaction. Yet it should not be thought that a good hypothesis is one that explains everything it is applied to. The smooth facility of Freudian and the older evolutionary formulae[1] exasperated those who were dissatisfied with them because, though explaining everything in general, they explained nothing in particular. Popper takes a scientifically realistic view: it is the daring, risky hypothesis, the hypothesis that might so easily not be true, that gives us special confidence if it stands up to critical examination.

3

How well does the 'hypothetico-deductive' scheme of thought measure up to the specifications of a good methodology, as I tried to lay them down at the end of Part II?

(1) A clear distinction is made between discovery and justification or proof,[2] now resolved into two separate and dissociable episodes of thought.

(2) The initiative for the kind of action that is distinctively scientific is held to come, not from the apprehension of 'facts', but from an imaginative preconception of what might be true.

(3) The hypothetico-deductive scheme provides a theory of special incentive. Our observations no longer range over the universe of observables: they are confined to those that have a bearing on the hypothesis under investigation.

(4) It allows also for the continual rectification or running adjustment of hypotheses by the process of negative feedback I shall mention again below.

(5) Error is simply explained – the fact that scientific research so very often goes wrong. Scientific error is now an ordinary part of human fallibility: we simply guess wrong, take a wrong view, form a mistaken opinion.

(6) Luck, unintelligible in inductive reasoning, now makes sense.

---

[1] For the shortcomings of older Darwinian formulae, see pp. 230–1 below.

[2] 'Proof' in the sense of probation, the act or action of testing or trial. 'Proof' in a logical context usually refers to a test of which the outcome is verification: proving a theorem is not just testing it, but testing it and finding it to be true.

The lucky accident fulfils a prior expectation, however vaguely for-mulated it may have been.[1]

(7) The hypothetico-deductive scheme gives due weight to the critical purposes of experimentation: we carry out experiments more often to discriminate between possibilities than to enlarge the stockpile of factual information.

So far all is well. Now let us turn to certain shortcomings, real or fancied, of the hypothetico-deductive scheme.

If it is a formal objection to classical inductivism that it sets no upper limit to the amount of factual information we should assemble, so also is it a defect of the hypothetico-deductive scheme that it sets no upper limit to the number of hypotheses we might propound to account for our observations. To substitute Whewell's system for Mill's is, on the face of it, to trade in an infinitude of irrelevant facts for an infinitude of inane hypotheses. Mill meant it as a criticism, not as a comment, when he said:[2]

An hypothesis being a mere supposition, there are no other limits to hypotheses than those of the human imagination; we may, if we please, imagine, by way of accounting for an effect, some cause of a kind utterly unknown, and acting according to a law altogether fictitious.

In real life, of course, just as the crudest inductive observations will always be limited by some unspoken criterion of relevance, so also the hypotheses that enter our minds will as a rule be plausible and not, as in theory they could be, idiotic. But this implies the existence of some internal censorship which restricts hypotheses to those that are not

[1] We scientists often miss things that are 'staring us in the face' because they do not enter into our conception of what might be true, or, alternatively, because of a mistaken belief that they could not be true (see p. 107 below, note 1). In our earlier work on immunological tolerance (*Philosophical Transactions of the Royal Society* B, **239**, 357–414, 1956), my colleagues R. E. Billingham and L. Brent and I completely missed the significance of observations which, rightly construed, would have led us to recognise an altogether new variant of the immunological response (the 'graft against host' reaction) which now plays a very important part in the theory of tissue transplantation. The 'facts' were before us, and if induction really worked we should not have been obliged to wait several years for their elucidation, which was hit upon independently by M. Simonsen and by Billing-ham and Brent themselves (see Billingham's Harvey Lecture on *The Biology of Graft-versus-Host Reactions* (New York, 1968)).
[2] *A System of Logic*, book III, chapter XIV, § 4.

absurd, and the internal circuitry of this process is quite unknown. The critical process in scientific reasoning is not therefore wholly logical in character, though it can be made to appear so when we look back upon a completed episode of thought.

A second objection is this: that although falsifiability is a logically conclusive process – if our inferences are false, the axioms from which we deduced them must be false also – we may yet be fallible in our imputation of fallibility. We could be mistaken in thinking that our observations falsified a hypothesis: the observations may themselves have been faulty, or may have been made against a background of misconceptions; or our experiments may have been ill-designed. The act of falsification is not immune to human error.[1]

My third point is comment, not criticism. There is nothing distinctively scientific about the hypothetico-deductive process. It is not even distinctively intellectual. It is merely a scientific context for a much more general stratagem that underlies almost all regulative processes or processes of continuous control, namely *feedback*, the control of performance by the consequences of the act performed. In the hypothetico-deductive scheme the inferences we draw from a hypothesis are, in a sense, its logical output. If they are true, the hypothesis need not be altered, but correction is obligatory if they are false. The continuous feedback from inference to hypothesis is implicit in Whewell's account of scientific method; he would not have dissented from the view that

---

[1] The force of this objection was impressed upon me by Ernest Nagel. I can remember forming the opinion, as a young research worker, that cells long maintained in culture outside the body undergo a transformation into the cancerous state, but dropped it because of the mistaken belief that transplantation experiments had already proved the idea untenable. My reasoning was that, if long-cultivated cells were in fact malignant (as many are now known to be), they should grow progressively, as malignant tumours do, when reimplanted into the body. In fact they did not do so, so the hypothesis was disproved. Unfortunately the act of disproof was itself erroneous: we now know that such a test could only have been valid if the cultivated cells contained no transplantation antigens not also present in the organism into which they were implanted; and even this test would have been unreliable if the cultivated cells had acquired new tumour-specific antigens during their growth outside the body. It is indeed true that 'disproof of a hypothesis is contingent on the stability of the theories employed in interpreting matters of fact, so that the refutation of a supposed explanation may be no more definitive than is its verification' (E. Nagel, in a review of *The Art of the Soluble*, *Encounter*, September 1967, pp. 68–70).

scientific behaviour can be classified as appropriately under cybernetics as under logic.[1]

The major defect of the hypothetico-deductive scheme, considered as a formulary of scientific behaviour, is its disavowal of any competence to speak about the generative act in scientific enquiry, 'having an idea', for this represents the imaginative or logically unscripted episode in scientific thinking, the part that lies outside logic. The objection is all the more grave because an imaginative or inspirational process enters into *all* scientific reasoning at every level: it is not confined to 'great' discoveries, as the more simple-minded inductivists have supposed.

Scientists are usually too proud or too shy to speak about creativity and 'creative imagination'; they feel it to be incompatible with their conception of themselves as 'men of facts' and rigorous inductive judgements. The role of creativity has always been acknowledged by inventors, because inventors are often simple unpretentious people who do not give themselves airs, whose education has not been dignified by courses on scientific method. Inventors speak unaffectedly about brain-waves and inspirations: and what, after all, is a mechanical invention if not a solid hypothesis, the literal embodiment of a belief or opinion of which mechanical working is the test?

Intuition takes many different forms in science and mathematics, though all forms of it have certain properties in common: the suddenness of their origin, the wholeness of the conception they embody, and the absence of conscious premeditation. The four examples I shall give are not meant to be exhaustive or mutually exclusive.

(a) *Deductive intuition*: perceiving logical implications instantly; seeing *at once* what follows from holding certain views. Inasmuch as deductive reasoning merely uncovers or brings to light what is implicit in our premises, a pedant might insist that deductive intuition is not a 'creative' process. To a perfect mind Pythagoras' Theorem (the one that startled Thomas Hobbes) would simply be a boring and repetitious way of inderlining a point which had already been made, much more

---

[1] 'Trial and error' will not do as a description of the process by which we devise and test hypotheses, for it carries the sense of random exploration, or of exploration according to a scheme ('Let's try the following possibilities in turn') which is not influenced by the testimony of prior mistakes.

compendiously, in Euclid's axioms.[1] It is indeed true that deduction owes its existence to the infirmity of our powers of reasoning: it cannot bring us news of the world, but (because our minds are indeed imperfect) it can bring us awareness.

(b) The form of intuition which, unless we are to abandon the word altogether, might as well be called *inductive*: thinking up or hitting on a hypothesis from which whatever we may wish to explain will follow logically.[2] This is the generative act in scientific discovery, the invention of a fragment of a possible world. 'Creativity' is a vague word, but it is in just such a context as this that we should choose to use it.

(c) The instant apprehension of analogy, that is, a real or apparent structural similarity between two or more schemes of ideas, regardless of what the ideas are about. No one word in common use describes this faculty in all its manifestations, but if I had to choose one word I should choose *wit* (compare the anecdote about Sydney Smith, p. 103 above).

(d) Most scientists cannot be classified as either experimentalists or theorists, because most of us are both, but we all recognise a distinction between the faculties that found superlative expression in, say, Michael Faraday on the one hand or James Clerk Maxwell on the other. For an experimentalist the most exciting and pleasing act in science is thinking up or thinking out an experiment which provides a really searching test of a hypothesis. We recognise the intuitive element in such a process when we speak of experimental flair or insight, but here too no one word in common speech stands for everything it should convey.

The analysis of creativity in all its forms is beyond the competence of any one accepted discipline. It requires a consortium of the talents: psychologists, biologists, philosophers, computer scientists, artists and poets will all expect to have their say. That 'creativity' is beyond analysis is a romantic illusion we must now outgrow. It cannot be learned perhaps, but it can certainly be encouraged and abetted. We can put ourselves in the way of having ideas, by reading and discussion

[1] See for example Hans Hahn's influential paper on 'Logic, Mathematics and Knowledge of Nature' (1933), reprinted in *Logical Positivism*, ed. A. J. Ayer (New York, 1959).
[2] It is a grievous blunder to speak, as so many do, of *deducing* hypotheses. Hypotheses are what we deduce things from.

and by acquiring the habit of reflection, guided by the familiar principle that we are not likely to find answers to questions not yet formulated in the mind. I am not offended by the idea that drugs may help us to formulate hypotheses, but I know of none which improves their quality, and I should hesitate to use a drug which did not enhance the critical faculty in proportion to the rate of accession of ideas.

4

The scheme of thought I have outlined in Part III explains the balance of faculties that should be cultivated in scientific research. Imaginativeness and a critical temper are both necessary at all times, but neither is sufficient. The most imaginative scientists are by no means the most effective; at their worst, uncensored, they are cranks. Nor are the most critically minded. The man notorious for his dismissive criticisms, strenuous in the pursuit of error, is often unproductive, as if he had scared himself out of his own wits – unless indeed his critical cast of mind was the consequence rather than the cause of his infertility.

The hypothetico–deductive system seems to me to give a reasonably lifelike picture of scientific enquiry, considered as a form of human behaviour. It makes science very human in its successes as well as in its fallibility. 'Let us look each other in the face,' said Nietzsche:[1] 'we are Hyperboreans: we know well enough how far outside the crowd we stand.' Nietzsche was speaking about philosophers, to be sure, and more particularly about himself; but philosophers have long since outgrown the hyperborean image, and (to whatever degree it may have been wished upon him) the scientist must outgrow it too. The scientific method is a potentiation of common sense, exercised with a specially firm determination not to persist in error if any exertion of hand or mind can deliver us from it. Like other exploratory processes, it can be resolved into a dialogue between fact and fancy, the actual and the possible; between what could be true and what is in fact the case. The purpose of scientific enquiry is not to compile an inventory of factual

[1] *The Antichrist*, trans. A. M. Ludovici (London, 1911). The Hyperboreans, inhabitants of a serene and timeless world lying beyond the roots of Boreas, the North Wind, seem to owe their existence to a clerical error: 'Hyperphoreans', I have been told, may be a sounder rendering of their original name, so they are doubly mythical creatures.

information, nor to build up a totalitarian world picture of Natural Laws in which every event that is not compulsory is forbidden. We should think of it rather as a logically articulated structure of justifiable beliefs about nature. It begins as a story about a Possible World – a story which we invent and criticise and modify as we go along, so that it ends by being, as nearly as we can make it, a story about real life.

APPENDIX
TO PAGE 83, NOTE I

The examples given in the text were chosen to make the point that many ideas belonging to a sociological level of discourse make no sense in biology and that many biological ideas make no sense in physics, but it is important not to forget that this restriction on the flow of thought works one way only. Nothing disqualifies the inclusion of physical or chemical propositions in the biological or social sciences. That gold should be used as a currency standard depends in part on its being incorruptible by rust and the ravages of lepidoptera, and to 'explain' this property we must investigate its physico-chemical properties or seek guidance from those who do so. The explanation of why an unexpectedly high proportion of Nigerians enjoy an inborn resistance to subtertian malaria turns on a specially detailed knowledge of the structure of the haemoglobin molecule. Examples of this kind are limitless. There is a sense in which the social sciences comprehend biology and make use of biological notions, and in which sociology is empirically and conceptually the richer subject. The same could be said of the biological sciences *vis-à-vis* physics and chemistry. Yet there is also a sense in which physics comprehends biology, and biology in its turn the social sciences, as the more general sciences comprehend the more particular.

These are exasperatingly vague statements, and they sound paradoxical: how can sociology be empirically and conceptually richer than biology if there is a sense in which biology comprehends it?

The only way I can make the case is by appealing to an analogy between the hierarchy of the empirical sciences and the hierarchy of classical geometries as it was envisaged by 'the greatest synthesist that geometry has ever known', Felix Klein. The reader must judge for himself whether or not the analogy is illuminating. In what follows I shall not strive after a rigour I am unqualified to achieve. (See J. L. Coolidge, *A History of Geometrical Methods* (Oxford, 1940); E. T. Bell, *The Development of Mathematics* (New York, 1940). Klein's great synthesis was adumbrated in 1872, and the best account of it is his own:

Elementary Mathematics from an Advanced Standpoint: Geometry, trans. E. R. Hedrick and C. Noble (London, 1939; 1st German ed., 1908). By 'classical' geometries I mean those in which geometric objects are represented as being contained (as by a vessel) in a space that is independent of them.)

In Klein's conception, a geometry is the invariant theory of a certain specified group of geometric operations, i.e. it is the class of statements describing the properties of geometric objects that remain *un*changed under the transformations to which they are subjected. A transformation may be thought of as a substitution of one set of points for another, keeping the same coordinate system; or, alternatively, as a substitution of one coordinate system for another, the points themselves being thought of as remaining unchanged. Whichever way we choose to think of it, the transformation can be described in a geometric language – we can speak of displacements, rotations, inversions etc. – or they may be defined by analytic (algebraic) formulae, 'mapping functions', which are rules for exchanging the new points (or new coordinates) for the old. (A 'group' of transformations is a set of which each member has an inverse, and which is such that the successive performance of any two transformations is itself a member of that set. 'Displacements' in space form a group, because any displacement (say from position A to B) has its inverse (B to A) which restores the *status quo*, and the product of two successive displacements is itself a displacement. In chess, pawns' moves have no inverse; knights' moves have an inverse, but the product of two knights' moves is not a knight's move. Neither forms a 'group'.)

In this scheme of codification, metric, Euclidian and affine geometries and topology may be said to form a hierarchy: we can pass from one to the other by progressively relaxing the conditions imposed by the rules of transformation, or (in the other direction) by making them progressively stricter. Metric geometry is the most highly restricted: the group of operations that defines it consists only of translations, rotations and inversions. The invariant theory of this group of operations is the richest in geometric concepts: it will contain a superabundance of theorems to do with isosceles triangles, regular polygons and degrees of curvature and angularity; it can make use of the idea of scalar *distance* also, for the distance between two points is invariant under the transformations of the metric group – transformations which conserve all properties associated with size and shape.

The Euclidian group of transformations is a little more permissive: symmetrical magnification is allowed and the concepts of size and metric distance therefore disappear, though the notions of (for example) square and circle are retained, and indeed all properties to do with *shape*, which is invariant.

Affine geometry is specified by a group of transformations which (in geometrical terms) allows for uniform magnification, but to different degrees in the three dimensions of space. The concepts of square and circle and size of angle are now meaningless, since the properties that define them are not invariant under transformation, but linearity and parallelism remain, and theorems to do with ellipses and parallelograms. Geometric objects which are Euclidian transforms of each other are described as 'similar'; when they are affine transforms of each other they are sometimes described as *homoeographic*. A mechanical or architectural drawing is a special kind of homoeograph of the object it represents. (Projective geometry, familiar from perspective drawing, includes affine geometry as a special case, but it does not belong in the hierarchy under discussion because the transformations that define it allow points to be 'carried to infinity'. In projective geometry, linearity remains, but not parallelism; and circle, ellipse etc. give way to the more general notion of a conic section.)

Topology is the most permissive of the four geometries, for nothing is required of the transformations that define it except that they should be continuous and should bring the transformed points into a one-to-one correspondence with the points they replace. A topological transformation may be represented geometrically by an arbitrary plastic deformation, such as a geometric figure would undergo if it were drawn upon a sheet of rubber which was thereupon stretched or twisted in any way that did not tear it. Figures related to each other by continuous plastic deformation retain the kind of primitive likeness that is described as *homoeomorphy*. Obviously all simple geometric notions have now lost their meaning, but certain very elementary properties remain, e.g. the order of points on a line, relationships of insideness and outsideness of closed figures, the 'sidedness' of surfaces (in these enlightened days all children are taught to play with Möbius strips).

It follows from the way in which they were derived that all the theorems of topology are 'true in' (and all topological concepts make sense in) affine geometry, that the theorems and concepts of affine geometry are part of Euclidian geometry, and so on. As we pass down the series – topology, affine geometry, Euclidian geometry, metric geometry – we may note that: (*a*) each geometry is a special case of its predecessor, i.e. is derived by imposing special restrictions upon or defining a sub-group within the one preceding it; (*b*) all theorems of one geometry are also theorems in its successors; (*c*) new concepts (e.g. of parallelism, circularity or shape) 'emerge' at each level which have no meaning and cannot be envisaged at an earlier level; and (*d*) there is a progressive enrichment in the number and variety of concepts and the particularity and degree of detail of the theorems.

My argument is that much the same kind of relationship holds between the elements of a hierarchy of the empirical sciences, taken in the order physics, chemistry, biology, sociology. The hierarchy is in this case compositional (inasmuch as people are made of molecules and societies of people), the rules of transformation are causal rather than algebraic, and the theorems are of empirical origin, but (*mutatis mutandis*) the relationships *a, b, c, d* are valid nevertheless. In the light of this interpretation the ideas of *reducibility* and *emergence* are no longer mystifying; nor is it self-contradictory to say that sociology comprehends biology even though it is, formally speaking, a 'special case' (in the sense that 'societies' are only a subclass of all possible systems of interaction between individuals, just as living organisms represent a subclass of all possible configurations or systems of interaction between molecules). It is a sociological truth as well as a physical truth that the atomic weight of sulphur is 32. The trouble about such a statement is not that it is false or meaningless in the social sciences, but that it is unimportant or dull. The same, however, would apply to a topological theorem in the context of metric geometry, e.g. 'Every right-angled isosceles triangle divides a plane into a part inside it and a part outside it.' 'What of it?' is the natural comment in either case.

This telegraphically condensed argument is set out in full in 'A Geometric Model of Reducibility and Emergence', in F. Y. Ayala and T. Dobzhansky (eds), *Studies in the Philosophy of Biology* (London, 1974). See also P. B. Medawar and J. S. Medawar, *The Life Science* (London, 1977), chapter 22.

HYPOTHESIS AND IMAGINATION

There is a mask of theory over the whole face of nature.

WILLIAM WHEWELL

I

IF AN educated layman were asked to set down his understanding of what goes on in the head when scientific discoveries are made and of what it is about a scientist that qualifies him to make them, his account of the matter might go something like this. A scientist is a man who has cultivated (if indeed he was not born with) the restless, analytical, problem-seeking, problem-solving temperament that marks his possession of a Scientific Mind. Science is an immensely prosperous and successful enterprise – as religion is not, nor economics (for example), nor philosophy itself – because it is the outcome of applying a certain sure and powerful method of discovery and proof to the investigation of natural phenomena: *The Scientific Method.* The scientific method is not deductive in character – it is a well-known fallacy to regard it as such – but it is rigorous nevertheless, and logically conclusive. Scientific laws are *in*ductive in origin. An episode of scientific discovery begins with the plain and unembroidered evidence of the senses – with innocent, unprejudiced observation, the exercise of which is one of the scientist's most precious and distinctive faculties – and a great mansion of natural law is slowly built upon it. Imagination kept within bounds may ornament a scientist's thought and intuition may bring it faster to its conclusions, but in a strictly formal sense neither is indispensable. Yet Newton was too severe upon hypotheses, for though there is indeed something *mere* about hypotheses, the best of them may look forward to a dignified middle age as Theories.[1]

A critic anxious to find fault might now raise a number of objections, among them these: (1) there is no such thing as a Scientific Mind;

[1] *Hypotheses non sequor*, runs an early draft of Newton's famous disclaimer, which we are to translate, as Whewell did, 'I feign no hypotheses': see I. Bernard Cohen, *Isis*, **51**, 589, 1960. Newton did, of course, use and propound hypotheses in the modern sense of that word; the unwholesome flavour which Newton found in the word is discussed below.

(2) there is no such thing as The Scientific Method; (3) the idea of naïve or innocent observation is philosophers' make-believe; (4) 'induction' in the wider sense that Mill gave it is a myth; and (5) the formulation of a natural 'law' always begins as an imaginative exploit, and without imagination scientific thought is barren. Finally (he might add) it is an unhappy usage that treats a hypothesis as an adolescent theory.

1   *There is no such thing as a Scientific Mind.* Scientists are people of very dissimilar temperaments doing different things in very different ways. Among scientists are collectors, classifiers and compulsive tidiers-up; many are detectives by temperament and many are explorers; some are artists and others artisans. There are poet-scientists and philosopher-scientists and even a few mystics. What sort of mind or temperament can all these people be supposed to have in common? *Obligative* scientists must be very rare, and most people who are in fact scientists could easily have been something else instead.

2   *There is no such thing as The Scientific Method* – as *the* scientific method, that is the point: there is no one rounded art or system of rules which stands to its subject-matter as logical syntax stands towards any particular instance of reasoning by deduction. 'An art of discovery is not possible,' wrote a former Master of Trinity; 'we can give no rules for the pursuit of truth which shall be universally and peremptorily applicable.' To many philosophers of science such an opinion must have seemed treasonable, and we can understand their unwillingness to accept a judgement that seems to put them out of business. The face-saving formula is that although there is indeed a Scientific Method, scientists observe its rules unconsciously and do not understand it in the sense of being able to put it clearly into words.

3   *The idea of naïve or innocent observation is philosophers' make-believe.* To good old British empiricists it has always seemed self-evident that the mind, uncorrupted by past experience, can passively accept the imprint of sensory information from the outside world and work it into complex notions; that the candid acceptance of sense-data is the elementary or generative act in the advancement of learning and the foundation of everything we are truly sure of.[1] Alas, unprejudiced

---

[1] The fundamental axiom of empiricism – *nihil in intellectu quod non prius in sensu* – is of course mistaken. Animals *inherit* information (for example, on how

observation is mythical too. In all sensation we pick and choose, interpret, seek and impose order, and devise and test hypotheses about what
we witness. Sense data are taken, not merely given: we *learn* to per-
ceive.[1] 'Why can't you draw what you see?' is the immemorial cry of
the teacher to the student looking down the microscope for the first
time at some quite unfamiliar preparation he is called upon to draw.
The teacher has forgotten, and the student himself will soon forget,
that what he sees conveys no information until he knows beforehand
the kind of thing he is expected to see. I cite more evidence on this
point below.

4 *Induction is a myth.* In donnish conversation we are not taken
aback when someone says he has 'deduced' something or has carried
out a deduction; but if he were to say he had *in*duced something or
other we should think him facetious if not a pompous idiot. So it is
with 'Laws': scientists do not profess to be trying to discover laws and
use the word itself only in conventional contexts (Hooke's Law,
Boyle's Law). (The actual usages of scientific speech are, as I shall
explain below, extremely revealing.) It is indeed a myth to suppose
that scientists actually carry out inductions or that a logical autopsy
upon a completed episode of scientific research reveals in it anything
that could be called an inductive structure of thought.

'Induction' in the wider sense that distinguishes it from perfect or
merely iterative induction (see below) is a word lacking the qualities
that would justify its retention in a professional vocabulary. It is seldom,
if ever, used in any sentence of which it is not itself the subject, and it
has no agreed meaning. *Finding* a meaning for induction has been a
philosophic pastime for more than a hundred years. Whewell used the
word, but with some feeling in later years that he might have dropped
it. 'There is really no such thing as a distinct process of induction,' said
Stanley Jevons; 'all inductive reasoning is but the inverse application
of deductive reasoning' – and this was what Whewell meant when he
said that induction and deduction went upstairs and downstairs on the

___

to build nests, or what to sing) in the form of a sort of chromosomal tape-
recording. This instinctual knowledge is not arrived at by association of ideas,
anyhow of sensory ideas received by the animal in its own lifetime.

[1] *The Organization of Behavior* by D. O. Hebb (New York, 1949), especially
p. 31.

same staircase. For Samuel Neil, however, 'induction' was confined to the act of testing a scientific conjecture or presupposition, and this was also C. S. Peirce's usage ('The operation of testing a hypothesis by experiment . . . I call induction').[1] Peirce accordingly uses the words *retroduction* or *abduction* to mean what Jevons called *in*duction. Nowadays the tendency is to use 'experimentation' to stand for the acts used in testing a hypothesis, leaving 'induction' as a vague word to signify all the various ways of travelling upstream of the flow of deductive inference. (Popper, of course, is for abandoning 'induction' altogether.)

The word *experiment* has also changed its meaning. When amateurs of the history of science attribute to Bacon the advocacy of the experimental method, they are often acting under the impression that Bacon used the word as we do. But a Baconian 'experiment' had the connotation that still persists in the French 'expérience' today: a Baconian experiment is a contrived experience or contrived happening as opposed to a natural experience or happening, for Bacon rightly supposed that common knowledge was not enough and that there was no relying upon luck of observation – upon 'the casual felicity of particular events'. The Philosophers of Mind took the same view: experiments were 'designed observations' intended 'to place nature in situations in which she never presents herself spontaneously to view, and to extort from her secrets over which she draws a veil to the eyes of others'.[2] Rubbing two sticks together to see what happens is an experiment in Bacon's sense; rubbing two sticks together to see if enough heat can be generated by friction to ignite them is an experiment in the modern sense. An experiment of the first kind leaves one with no answer to the

[1] Here and hereafter I quote from the following works of the authors cited under heading 4 in the text:

The Philosophy of the Inductive Sciences, by William Whewell, in 2 vols, 2nd ed. (London, 1847; 1st ed., 1840).

The Principles of Science, by W. Stanley Jevons, 2nd ed., revised (London, 1877; 1st ed., 1873).

The Art of Reasoning, by Samuel Neil: twenty articles in successive issues of the first two vols of the *British Controversialist* (1850, 1851), of which Neil was editor; particularly vol. 2, no. 11.

Collected Papers of C. S. Peirce, ed. C. Hartshorne & P. Weiss, vol. 2, *Elements of Logic* (Cambridge, Mass., 1932); vol. 6, *Scientific Metaphysics* (1935).

[2] *Elements of the Philosophy of the Human Mind*, by Dugald Stewart, 2nd ed., 2 vols (London, 1802, 1816; 1st ed., 1792, 1814).

question (a 'good' question: see below), 'Why on earth are you rubbing those two sticks together?'

I shall refer later to the changing connotation of 'hypothesis', a word that has grown in stature as 'induction' has declined.

The concept of induction was entrenched into scientific method-ology through the formidable advocacy of John Stuart Mill. Mill, said John Venn in 1907,[1] had 'dominated the thought and study of intelligent students to an extent which many will find it hard to realise at the present day'; yet he could still take a general familiarity with Mill's views for granted, in spite of having recorded as far back as 1889 'a broadening current of dissatisfaction' on the part of physicists which had 'mostly taken the form of an ill-concealed or openly avowed con-tempt of the logical treatment of Induction'. It is, however, the indiffer-ence rather than the hostility of critically-minded scientists that has allowed the myth of induction to persist – combined, I believe, with the great earnestness and sincerity of Mill himself; for Mill believed, as so many good people believe today, that if only we could formulate and master The Scientific Method many of the vexed problems of modern society would vanish before its use.

Mill's was, of course, Induction in the strong, imperfect, or open-ended sense. 'Induction', said Mill ('that great mental operation'), 'is a process of inference; it proceeds from the known to the unknown; and any operation involving no inference, any process in which what seems the conclusion is no wider than the premises from which it is drawn, does not fall within the meaning of the term.' That would be very well if he had not also said that induction was an exact and logi-cally rigorous process, capable of doing for empirical reasoning what logical syntax does for the process of deduction: 'The business of in-ductive logic is to provide rules and models (such as the syllogism and its rules are for ratiocination) to which, if inductive arguments con-form, those arguments are conclusive, and not otherwise.'[2]

There seems no point in mulling over the logical errors of Mill's *System*, for they are now common knowledge – for example, his failure to distinguish between the methodologies of discovery and of

---

[1] *The Principles of Empirical or Inductive Logic*, 2nd ed. (London, 1907; 1st ed., 1889).

[2] *A System of Logic*, by John Stuart Mill, 8th ed. (London, 1872; 1st ed., 1843).

proof (though Whewell had insisted on the distinction), and the circularity of his attempt to justify that 'ultimate syllogism' which had 'for its major premise the principle or axiom of the uniformity of the course of nature'. But one may yet be surprised by how little he understood the methodological functions of hypotheses, and by the hopeless ambition embodied in his belief that it was possible merely by taking thought to arrive with certainty at the truth of general statements containing more information than the sum of their known instances.

The current of informed opinion was already flowing in the other direction. The probationary character of scientific law is implicit in all of Whewell, and long before him George Campbell, in his influential and widely read *Philosophy of Rhetoric* (1776), had said of inductive generalisation that there 'may be in every step, and commonly is, less certainty than in the preceding; *but in no instance whatever can there be more*' (my italics). 'No hypothesis', said Dugald Stewart, 'can completely exclude the possibility of exceptions or limitations hitherto undiscovered.' By the latter half of the nineteenth century the point had become commonplace. 'No inductive conclusions are more than probable,' said Jevons: 'we never escape the risk of error altogether.' Venn took pains to emphasise his belief 'that no ultimate objective certainty, such as Mill for instance seemed to attribute to the results of induction, is attainable by any exercise of the human reason'. 'The conclusions of science make no pretence to being more than probable,' wrote C. S. Peirce.

The logical status of deduction and syllogistic reasoning had not been seriously in question since the days of Bacon. Syllogistic reasoning (an 'unnatural art', Campbell had called it, and others 'futile' or 'puerile') was indeed a logically conclusive process, but that was because it merely 'expands and unfolds', merely brings to light and makes explicit the information lying more or less deeply hidden in the premises out of which it flows. Deduction makes known to us only what the infirmity of our powers of reasoning has so far left concealed. The case had been well put by Archbishop Whately,[1] and Mill accepted it; and so the peculiar and distinctive role of deduction in scientific reasoning came to be overlooked. Convinced nevertheless that Science had come upon irrefragable general truths by some process other than deduction,

[1] *Elements of Logic*, by Richard Whately, 9th ed. (London, 1848; 1st ed., 1826).

Mill had no alternative but to put his faith in induction – to believe in the existence of a valid inductive process even if his own account of it should prove faulty or incomplete.

What about Baconian induction – the painstaking assembly and classification of natural and elicited (experimental) facts of which Jevons said that it reduced the methodology of science to a kind of bookkeeping? By the sixth edition of the *Origin* in 1876, Darwin had convinced himself that he had been a good Baconian, but his correspondence tells a different story. Darwin's status as the culture-hero of induction – the great but deeply humble scientist listening attentively to Nature's lessons from her own lips – has now to be reconciled with evidence that he had the germ of the idea of natural selection before ever he had read Malthus.[1]

It is Karl Pearson whose scientific practice and theoretical professions earn him the right to be called a true Baconian. 'The classification of facts', he wrote in *The Grammar of Science*, and

the recognition of their sequence and relative significance is the function of science . . . let us be quite sure that whenever we come across a conclusion in a scientific work which does not flow from the classification of facts, or which is not directly stated by the author to be an assumption, then we are dealing with bad science.[2]

Poor Pearson! His punishment was to have practised what he preached, and his general theory of heredity, of genuinely inductive origin, was in principle quite erroneous.

I have given here the conventional view of Bacon's methodology, and shall return later to the claim made on his behalf by Coleridge and others that he was fully aware of the methodological value of hypotheses.

5   *The formulation of a natural law begins as an imaginative exploit and*

[1] *Charles Darwin*, by Gavin de Beer (London, 1963), p. 98. In his Autobiography Darwin once declared that he could not resist forming a hypothesis on every subject, and his letters to Henry Fawcett and to H. W. Bates are very revealing: *More Letters of Charles Darwin*, ed. F. Darwin and A. C. Seward (London, 1903), pp. 176, 195. See also p. 80 above, note 2.

For the origin of the idea of Natural Selection, see also L. Eisely in *Daedalus*, Summer 1965, pp. 588–602.

[2] *The Grammar of Science*, by Karl Pearson, 3rd ed. (London, 1911; 1st ed., 1892).

imagination is a faculty essential to the scientist's task. Most words of the philosopher's vocabulary, including 'philosopher' itself, have changed their usages over the past few hundred years.[1] 'Hypothesis' is no exception. In a modern professional vocabulary a hypothesis is an imaginative preconception of *what might be true* in the form of a declaration with verifiable deductive consequences. It no longer tows 'gratuitous', 'mere', or 'wild' behind it, and the pejorative usage ('Evolution is a mere hypothesis', 'It is only a hypothesis that smoking causes lung cancer') is one of the outward signs of little learning. But in the days of Travellers' Tales and Marvels, when (as John Gregory contemptuously remarked)[2] philosophers were more interested in animals with two heads than in animals with one, 'hypothesis' carried very strongly the connotation of the wantonly fanciful and above all the gratuitous; nor was there any thought that a hypothesis need do more than explain the phenomena it was expressly formulated to explain. The element of *responsibility* that goes with the formulation of a hypothesis today was altogether lacking. Thomas Burnet's *Sacred Theory of the Earth* (1684–90) is a case in point – a romantic and absurd cosmology using the word 'hypothesis' in just the sense that Newton repudiated. 'Men of short thoughts and little meditation,' Burnet says in self-defence, 'call such theories as these, Philosophick Romances.' But, he says.

there is no surer mark of a good Hypothesis, than when it doth not only hit luckily in one or two particulars but answers all that it is to be applied to, and is adequate to Nature in her whole extent.

But how fully or easily soever these things may answer Nature, you will say, it may be, that all this is but an Hypothesis; that is, a kind of fiction or supposition that things were so and so at first, and by the coherence and agreement of the Effects with such a supposition, you would argue and prove that this is so. This I confess is true, this is the method, and if we would know anything in Nature further than our senses go, we can know it no otherwise than by an Hypothesis . . . and if that Hypothesis be easie and intelligible, and answers all the phaenomena . . . you have done as much as a Philosopher or as Humane reason can do.

[1] e.g. 'science', 'art', 'pure science', 'applied science', 'analysis', 'synthesis', 'experiment'; and, of course, notoriously, words like 'genius', 'creation', 'enthusiasm'.

[2] *On the Duties and Qualifications of a Physician*, new ed. (London, 1820; 1st ed., 1772).

Burnet's reasoning thus ends at the very point at which scientific reasoning begins. He did not seem to realise that his hypothesis about what the Earth was like before the flood and what it would be like after the Fire was but one among a virtual infinitude of hypotheses, and that he was under a moral obligation to find out if his were preferable to any other.

Burnet's preposterous speculations were expounded in prose that earned him immortality; because they were not, many philosophic romances that must have been known to Newton are now forgotten. Thomas Reid shall be allowed to sum the prevailing situation up.[1] 'It is genius,' he says, 'and not the want of it, that adulterates philosophy, and fills it with error and false theory. A creative imagination disdains the mean offices of digging for a foundation,' leaving these servile employments to scientific drudges. 'The world has been so long be-fooled by hypotheses in all parts of philosophy,' that we must learn 'to treat them with just contempt, as the reveries of vain and fanciful men.' Newton *could* have invented a hypothesis to account for gravitation, but 'his philosophy was of another complexion', for Newton had been 'taught by Lord Bacon to despise hypotheses as fictions of human fancy'.

Because Newton is cast as the hero of every scientific methodology of the past two hundred years, philosophers who attached great importance to hypotheses felt it their duty to explain away Newton's famous and profoundly influential disavowal. Stanley Jevons was so sure that Newton had practised what is now often called the 'hypo-thetico-deductive' method that he was inclined to think *hypotheses non fingo* ironical. But for two hundred years after Newton no one could advocate the use of hypotheses without an uneasy backward glance. Dugald Stewart said that an 'indiscriminate zeal against hypotheses' had been 'much encouraged by the strong and decided terms in which, on various occasions, they are reprobated by Newton'. 'Newton appears to have had a horror of the term *hypothesis*,' said William Whewell. Sir John Herschel spoke up in favour of hypotheses.[2] Samuel Neil in 1851 deplored the 'widely prevalent prejudice in the

[1] *Essays on the Intellectual Powers of Man*, 1st ed., 1785. In *The Works of Thomas Reid, D.D.*, ed. W. Hamilton, 4th ed. (London, 1854).
[2] *Discourse on the Study of Natural Philosophy* (London, 1831).

present age against hypotheses', and Thomas Henry Huxley had felt obliged to say, 'Do not allow yourselves to be misled by the common notion that a hypothesis is untrustworthy merely because it is a hypothesis.' Even George Henry Lewes found himself unable to propound his fairly sensible views on hypotheses without much prevarication and pursing of the lips.[1]

Where does Mill stand? Modern philosophers who are for various reasons 'pro-Mill' can of course find him a devotee of hypotheses. Hypotheses, Mill will be found to say, provide 'temporary aid', even 'large temporary assistance' ('temporary' because hypotheses are the larval forms of theories); hypotheses are valuable because they suggest observations and experiments, and in this respect they are indeed indispensable. However, all this had been said before, repeatedly: some instances I shall cite later. In his less conventional utterances on hypotheses Mill betrayed that he had no deep understanding of what is now thought to be their distinctive methodological function. He feared that people who used hypotheses did so under the impression that a hypothesis must be true if the inferences drawn from it were in accordance with the facts. Later therefore he says: 'It seems to be thought that an hypothesis . . . is entitled to a more favourable reception, if besides accounting for all the facts previously known, it has led to the anticipation and prediction of others which experience afterwards verified.'[2] This, he says, is 'well calculated to impress the uninformed', but will not impress thinkers 'of any degree of sobriety'.

Mill feared the imaginative element in hypotheses: 'a hypothesis being a mere supposition, there are no other limits to hypotheses than those of the human imagination'. These are Reid's fears, preying on Mill at a time when good reasons for feeling fearful had largely disappeared. Today we think the imaginative element in science one of its chief glories. Even Karl Pearson recognised it as a motive force in *great* discoveries, but, of course, 'imagination must not replace reason in the deduction of relation and law from classified facts'. (The belief that great discoveries and little everyday discoveries have quite different methodological origins betrays the amateur. Whewell, the professional,

---

[1] *Problems of Life and Mind*, 4th ed. (London, 1883; 1st ed. 1873), esp. pp. 296, 316–17.

[2] *A System of Logic*, book III, chapter 14, § 6.

insisted that the bold use of the imagination was the rule in scientific discovery, not the exception: see below.) All the same, the idea that hypotheses arose by mere conjecture, by guesswork, was thought undignified. Whewell had called good hypotheses 'happy guesses', though elsewhere, as if the occasion called for something more formal, he spoke of 'felicitous strokes of inventive talent'. But philosophers like Venn did not take to it: 'it is . . . scarcely an exaggeration of Whewell's account of the inductive process to say of it, as in fact has been said, that it simply resolves itself into making guesses'.

It is the word that is at fault, not the conception. To say that Einstein formulated a theory of relativity by guesswork is on all fours with saying that Wordsworth wrote rhymes and Mozart tuneful music. It is cheeky where something grave is called for.

2

I now turn to consider the history during the eighteenth and nineteenth centuries of some of the central ideas of the hypothetico-deductive scheme of scientific reasoning, confining myself, as hitherto, almost wholly to English and Scottish philosophers and the tradition of thought they embody. Among these ideas are:

(1) the uncertainty of all 'inductive' reasoning and the probationary status of hypotheses;

(2) the role of the hypothesis in starting enquiry and giving it direction, so confining the domain of observation to something smaller than the whole universe of observables;

(3) the asymmetry of proof and disproof: only disproof is logically conclusive;

(4) the obligation to put a hypothesis to the test.

1   I have already mentioned a number of earlier opinions on the inconclusiveness of scientific reasoning (above, p. 120).

2   It is our imaginative preconception of what might be true that gives us an incentive to seek the truth and a clue to where we might find it. 'In every useful experiment,' said John Gregory, writing in 1772, 'there must be some point in view, some anticipation of a principle to be established or rejected.' Such anticipations, he went on

to say, are *hypotheses*: people were suspicious of hypotheses because they did not fully understand their purpose, but without them 'there could not be useful observation, nor experiment, nor arrangement, because there would be no motive or principle in the mind to form them'. Dugald Stewart quoted passages expressing the same opinion in the writings of Boscovich, Robert Hooke, and Stephen Hales[1] – scientists all three. But on this point Coleridge sweeps everyone else aside.[2] In every advance of science, he assures us, 'a previous act and conception of the mind . . . an *initiative* is indispensably necessary', for when it comes to founding a theory on generalisation, 'what shall determine the mind to one point rather than another; within what limits, and from what number of individuals shall the generalization be made? *The theory must still require a prior theory for its own legitimate construction*' (my italics). Coleridge (like Stewart and later Neil) managed to convince himself of the great Francis Bacon's full awareness of the need for an 'intellectual or mental *initiative*' as the 'motive and guide of every philosophical experiment . . . namely, some well-grounded purpose, some distinct impression of the probable results, some self-consistent anticipation . . . which he affirms to be the prior *half* of the knowledge sought, *dimidium scientiae*'. The passage all three quote[3] as evidence for this interpretation is, in my reading of it, too slight to carry so great a weight of meaning.

3 Many philosophers in the older and the newer senses have spoken of the value of false hypotheses, and Stewart particularly commends the opinions of Boscovich ('the slightest of whose logical hints are entitled to particular attention'). Boscovich had said that by means of hypotheses

we are enabled to supply the defects of our *data*, and to conjecture or divine the path to truth; always ready to abandon our hypothesis, when found to involve consequences inconsistent with fact. And, indeed, in most cases, I conceive this to be the method best adapted to physics; a science in which . . .

[1] See Stephen Hale's Preface to his *Statical Essays*, 4th ed. (London, 1769; 1st ed., 1727) and a number of passages in Robert Hooke's *Posthumous Works* (London, 1705). For Boscovich, see p. 127 below, note 1.

[2] *On Method*, by Samuel Taylor Coleridge, 3rd ed. (London, 1849; 1st ed., 1818). See also p. 79 above, note 1.

[3] *De augmentis scientiarum*, trans. Gilbert Wats (London, 1674), book 5, chapter 3, II.

legitimate theories are generally the slow result of disappointed essays, and of errors which have led the way to their own detection.[1]

This is all right as far as it goes, but what one will not find so easily is a premonition of one of the strongest ideas in Popper's methodology, that the only act which the scientist can perform with complete logical certainty is the repudiation of what is false. It is *falsification* that has the logical stature attributed by the logical positivists to verification. 'Every experiment may be said to exist only in order to give the facts a chance of disproving the null hypothesis.'[2] The asymmetry of proof, considered as a point of logic, is of course very elementary, and it is merely slovenly or simple-minded to suppose that hypotheses are proved true if they lead to true conclusions. No logician of science has ever done so. Whewell certainly realised that refutation was methodologically a strong procedure, stronger than confirmation, an opinion that comes out more clearly in his aphorisms than in the body of the text:

(ix) The truth of tentative hypotheses must be tested by their application to facts. The discoverer must be ready, carefully to try his hypotheses in this manner . . . and to reject them if they will not bear the test.

(x) The process of scientific discovery is cautious and rigorous, not by abstaining from hypotheses, but by rigorously comparing hypotheses with facts, and by resolutely rejecting all which the comparison does not confirm.

These opinions shocked Mill. Dr Whewell's system, he complained, did not recognise 'any necessity for proof': 'If, after assuming an hypothesis and carefully collating it with facts, nothing is brought to light inconsistent with it, that is, if experience does not *dis*prove it, he is content; at least until a simpler hypothesis, equally consistent with experience, presents itself.' To Mill this attitude of Whewell's betrayed 'a radical misconception of the nature of the evidence of physical truths'. No wonder Venn said that a person who read both Mill and Whewell would find it hard to believe that they were discussing the same subject! The verdict must go to Whewell, 'whose acquaintance with the processes of thought of science', said Peirce, 'was incomparably greater than Mill's'.

[1] Dugald Stewart's translation of the footnotes on pp. 211–12 of Boscovich's *De solis ac lunae defectibus* (London, 1760).

[2] *The Design of Experiments*, by R. A. Fisher (Edinburgh, 1935).

4   The formulation of a hypothesis carries with it an obligation to test it as rigorously as we can command skills to do so. There was no sign of any such sense of obligation in Burnet's *Sacred Theory*: to explain the phenomena it was designed to explain was judged evidence enough. It satisfied curiosity in much the same way as a mother's desperately *ad hoc* answers satisfy the insistent questioning of a child. The child is not interested in the content of the answer: he asks as if he were under an instinctual compulsion to do so, and the act of answering completes a sort of ritual of exploration. But when curiosity is satisfied it is discharged: formulation of a hypothesis may act as a deterrent rather than as a stimulus to enquiry – a danger the earlier critics of the use of hypotheses were fully aware of.

Even the more sophisticated authors of Philosophick Romances did not seem to realise that any one set of phenomena could be explained by many hypotheses other than the one they fancied. It seems a strange blindness, but I think that Dugald Stewart in a finely reasoned passage got to the bottom of it. It was a favourite conceit in eighteenth-century philosophising – Stewart found it in Boscovich, Le Sage, D'Alembert, Gravesande and Hartley – that natural philosophy is, in David Hartley's words, 'the art of *decyphering* the Mysteries of Nature . . . so that . . . every Theory which can explain all the Phaenomena, has all the same Evidence in its favour, that it is possible the Key of a Cypher can have from its explaining that Cypher'.[1] Stewart found the analogy inept for many reasons, the chief being that whereas a cypher has one key, a unique solution, physical hypotheses seldom, if ever, 'afford the *only* way of explaining the phenomena to which they are applied'.

It is all very well to say that we are under a permanent obligation to test hypotheses, but, as Peirce said, 'there are some hypotheses which are of such a nature that they can never be tested at all. Whether such hypotheses ought to be entertained at all, and if so in what sense, is a serious question.' Certainly the logical positivists took the question very seriously indeed, and Popper has done so too, but I do not recollect its having been a live issue before Peirce.

---

[1] *Observations on Man*, vol. 1 (London, 1749), pp. 15–16.

3

Let me now set out the gist of the hypothetico-deductive system as it might be formulated today. ('Gist' is the right word, for there is no question of its providing an abstract formal framework which becomes a concrete example of scientific reasoning when we fill in the blanks.) First, there is a clear distinction between the acts of mind involved in discovery and in proof. The generative or elementary act in discovery is 'having an idea' or proposing a hypothesis. Although one can put oneself in the right frame of mind for having ideas and can abet the process, the process itself is outside logic and cannot be made the subject of logical rules. Hypotheses must be tested, that is criticised. These tests take the form of finding out whether or not the deductive consequences of the hypothesis or systems of hypotheses are statements that correspond to reality. As the very least we expect of a hypothesis is that it should account for the phenomena already before us, its 'extra-mural' implications, its predictions about what is not yet known to be the case, are of special and perhaps crucial importance. If the predictions are false, the hypothesis is wrong or in need of modification; if they are true, we gain confidence in it, and can, so to speak, enter it for a higher examination; but if it is of such a kind that it cannot be falsified even in principle, then the hypothesis belongs to some realm of discourse other than science. Certainty can be aspired to, but a 'rightness' that lies beyond the possibility of future criticism cannot be achieved by any scientific theory. There is no place for apodictic certainty in science.

The first strongly reasoned and fully argued exposition of a hypothetico-deductive system is unquestionably Karl Popper's. Quite a large part of it had been propounded at the level of learned discourse rather than of critical analysis by William Whewell, FRS, Master of Trinity College, Cambridge, in 1840. Whewell is never heard of nowadays outside the ranks of historians of science: if one mentions his name one may be asked to spell it. But his reputation in his day was formidable. Whewell wrote upon ethics, hydrostatics, political economy, astronomy, verse composition, terminology, the Platonic dialogues, mechanics, geology, and the History and Philosophy of the Inductive Sciences. He was the first *scientist*, I believe, to express a lengthy and carefully thought out opinion on the nature of scientific discovery, and in a sense the first scientist of any description, for he invented the word itself.

There are many inadequacies in Whewell, but the spirit is right. No general statement, he said, not even the simplest iterative generalisation, can arise merely from the conjunction of raw data. The mind always makes some imaginative contribution of its own, always 'superinduces' some idea upon the bare facts. A hypothesis is an explanatory conjecture giving one of many possible explanations that might meet the case.

A facility in devising hypotheses, therefore, is so far from being a fault in the intellectual character of a discoverer, that it is, in truth, a faculty indispensable to his task.

To form hypotheses, and then to employ much labour and skill in refuting, if they do not succeed in establishing them, is a part of the usual process of inventive minds. Such a proceeding belongs to the rule of the genius of discovery, rather than (as has often been taught in modern times) to the *exception*.

Yet it is indispensably necessary for the discoverer to demand of his hypotheses 'an agreement with facts such as will withstand the most patient and rigid inquiry', and, if they are found wanting, to turn them resolutely down:

Since the discoverer has thus constantly to work his way onwards by means of hypotheses, false and true, it is highly important for him to possess talents and means for rapidly *testing* each supposition as it offers itself.

The hypotheses which we accept ought to explain phenomena which we have observed. But they ought to do more than this: our hypotheses ought to *foretell* phenomena which have not yet been observed [but which are] of the same kind as those which the hypothesis was invented to explain.

Whewell did not believe that a scientist acquired factual information by passive attention to the evidence of his senses; the idea of 'naïve' or 'innocent' observation (see above, p. 116) he rejected altogether: 'Facts cannot be observed as Facts except in virtue of the Conceptions which the observer himself unconsciously supplies.' The distinction between fact and theory was by no means as distinct as people were accustomed to believe: 'There is a mask of theory over the whole face of nature.' Strictly speaking, no scientific discovery can be made by accident. What Whewell has to say on Man as the Interpreter of Nature[1] is a suitable prolegomenon to Popper's famous lecture 'On the Sources of Knowledge and of Ignorance'.

The account of scientific method which became recognised as the official alternative and rival to Mill's was not Whewell's but Stanley Jevons's. Jevons is not as fresh as Whewell nor so boldly original; we may think he should have acknowledged Whewell more often than he did. Jevons gave it as his 'very deliberate opinion' that 'many of Mill's innovations in logical science . . . are entirely groundless and false'. As to Bacon, he took the 'extreme view of holding that Francis Bacon . . . had no correct notions as to the logical method by which from particular facts we educe laws of nature'. Jevons endeavoured to show that 'hypothetical anticipation of nature is an essential part of inductive inquiry', the method 'which has led to all the great triumphs of scientific research'. Even in the 'apparently passive observation of a phenomenon' our attention should be 'guided by theoretical anticipations'.

The three essential stages in the process which he continued with deliberate vagueness to call 'induction' were, using his own words,

(a) Framing some hypothesis as to the character of the general law.

(b) Deducing consequences from that law.

(c) Observing whether the consequences agree with the particular facts under consideration.

Hypothesis is always employed, he says, consciously or unconsciously.

This account of the matter had come to be pretty widely agreed upon during the second half of the nineteenth century. We shall find it in Neil and Adamson[2] and very clearly in Peirce. (Venn, in spite of his reputation, I find disappointing.) There are many premonitions of the hypothetico-deductive method in the eighteenth century and even earlier, particularly in the writing of scientists. The clearest known to

---

[1] In particular see book 1, chapter 2, §§ 9, 10 (2nd ed.).

[2] Robert Adamson in his article 'Bacon, Francis' in the 9th ed. of the *Encyclopaedia Britannica* (Edinburgh, 1875). See also Augustus de Morgan's *A Budget of Paradoxes* (London, 1872): 'Modern discoveries have not been made by large collections of facts . . . A few facts have suggested an *hypothesis*, which means a *supposition*, proper to explain them. The necessary results of this supposition are worked out, and then, and not till then, other facts are examined to see if these ulterior results are found in nature. The trial of the hypothesis is the *special object* . . . Wrong hypotheses, rightly worked from, have produced more useful results than unguided observation.'

me is Dugald Stewart's, a point worth making because of the dismissive and totally erroneous opinion that his philosophy is simply a reproduction of his master's, Thomas Reid's, voice. In answer to Reid's rhetorical challenge to name any advance in science which had arisen by the use of a hypothetical method, Stewart thought it sufficient to mention the theory of gravitation and the Copernican system.

Stewart believed that most discoveries in science had grown out of hypothetical reasoning:

It is by reasoning synthetically from the hypothesis, and comparing the deductions with observation and experiment, that the cautious inquirer is gradually led, either to correct it in such a manner as to reconcile it with facts, or finally to abandon it as an unfounded conjecture. Even in this latter case, an approach is made to the truth in the way of *exclusion* . . .

Stewart's own analysis of the use of those tiresome adjectives *synthetic* and *analytic* shows he is here using 'synthetically' in the sense of 'deductively'.

4

A scientific methodology, being itself a theory about the conduct of scientific enquiry, must have grown out of an attempt to find out exactly what scientists do or ought to do. The methodology should therefore be measured against scientific practice to give us confidence in its worth. Unfortunately, this honest ambition is fraught with logical perils. If we assume for the sake of argument that the methodology is unsound, then so also will be our test of its validity. If we assume it to be sound, then there is no point in submitting it to test, for the test could not invalidate it. These difficulties I shall surmount by disregarding them entirely.

What scientists *do* has never been the subject of a scientific, that is, an ethological enquiry. It is no use looking to scientific 'papers', for they not merely conceal but actively misrepresent the reasoning that goes into the work they describe.[1] If scientific papers are to be accepted for

[1] See Popper's 'Science: Problems, Aims, Responsibilities', in *Federation Proceedings* (Federation of American Societies for Experimental Biology), 22, 961–72, 1963, and my own broadcast 'Is the Scientific Paper a Fraud?' in the *Listener*, 12 September 1963. This broadcast was followed by a correspondence (issues of 26 September and 10 October) illustrating the style of thought that makes scientists treat the 'philosophy of science' with exasperated contempt.

publication, they must be written in the inductive style. The spirit of John Stuart Mill glares out of the eyes of every editor of a Learned Journal.

Nor is it much use listening to accounts of what scientists *say* they do, for their opinions vary widely enough to accommodate almost any methodological hypothesis we may care to devise. Only unstudied evidence will do – and that means listening at a keyhole. Here are some turns of speech we may hear in a biological laboratory:

'What gave you the idea of trying . . .?'
'I'm taking the view that the underlying mechanism is . . .'
'What happens if you assume that . . .?'
'Actually, your results can be accounted for on a quite different hypothesis.'
'It follows from what you are saying that if . . ., then . . .'
'Is that actually the case?'
'That's a good question.' [i.e. a question about a true weakness, insufficiency or ambiguity]
'That result squared with my hypothesis.'
'So obviously that idea was out.'
'At the moment I don't see any way of eliminating that possibility.'
'My results don't make a story yet.'
'I'm still at the stage of trying to find out if there is anything to be explained.'
'Obviously a great deal more work has got to be done before . . .'
'I don't seem to be getting anywhere.'

Scientific thought has already reached a pretty sophisticated professional level before it finds expression in language such as this. This is not the language of induction. It does not suggest that scientists are hunting for facts, still less that they are busy formulating 'laws'. Scientists are building explanatory structures, *telling stories* which are scrupulously tested to see if they are stories about real life.

It has been a tradition among philosophers that we should look to the physical sciences and to simple, lofty discoveries if we are to see the Scientific Method at work in its most easily intelligible form. I question this opinion. The simplicity of great discoveries is often a measure of how far they have travelled from their beginnings. Let a biologist have a turn. Here is Claude Bernard, writing just one hundred years ago: 'A hypothesis is . . . the obligatory starting point of all experimental reasoning. Without it no investigation would be possible, and one would learn nothing: one could only pile up barren observations. To experiment without a preconceived idea is to wander aimlessly.' Indeed, 'Those who have condemned the use of hypotheses and preconceived ideas in the experimental method have made the mistake of confusing the contriving of the experiment with the verification of its results.'[1] Over and over again Bernard insists that hypotheses must be of such a kind that they can be tested, that one should go out of one's way to find means of refuting them, and that 'if one proposes a hypothesis which experience cannot verify, one abandons the experimental method'. Claude Bernard is most distinctive and at his best in his insistence on the critical method, on the virtue and necessity of Doubt.

When propounding a general theory in science, the one thing one can be sure of is that, in the strict sense, such theories are mistaken. They are only partial and provisional truths which are necessary . . . to carry the investigation forward; they represent only the current state of our understanding and are bound to be modified by the growth of science . . .

This is powerful evidence, for Claude Bernard, in creating experimental physiology, did indeed put scientific medicine on a new foundation. His philosophy *worked*.

In real life the imaginative and critical acts that unite to form the hypothetico-deductive method alternate so rapidly, at least in the earlier stages of constructing a theory, that they are not spelled out in thought. The 'process of invention, trial, and acceptance or rejection of the hypothesis goes on so rapidly,' said Whewell, 'that we cannot trace it in its successive steps'. What then is the point of asking ourselves where the initiative comes from, the observation or the idea? Is it not as pointless as asking which came first, the chicken or the egg?

But this is not a pointless question: it matters terribly which came first: scientific dynasties have been overthrown by giving the wrong answer! It matters no less in methodology: we may collect and classify facts, we may marvel at curiosities and idly wonder what accounts for them, but the activity that is characteristically scientific begins with an

[1] op. cit. (p. 73 above, note 1).

explanatory conjecture which at once becomes the subject of an energetic critical analysis. It is an instance of a far more general stratagem that underlies every enlargement of general understanding and every new solution of the problem of finding our way about the world. The regulation and control of hypotheses is more usefully described as a *cybernetic* than as a logical process: the adjustment and reformulation of hypotheses through an examination of their deductive consequences is simply another setting for the ubiquitous phenomenon of negative feedback. The purely logical element in scientific discovery is a comparatively small one, and the idea of a *logic* of scientific discovery is acceptable only in an older and wider use of 'logic' than is current among formal logicians today.

The weakness of the hypothetico-deductive system, in so far as it might profess to offer a complete account of the scientific process, lies in its disclaiming any power to explain how hypotheses come into being. By 'inspiration', surely: by the 'spontaneous conjectures of instinctive reasoning', said Peirce: but what then? It has often been suggested that the act of creation is the same in the arts as it is in science:[1] certainly 'having an idea' – the formulation of a hypothesis – resembles other forms of inspirational activity in the circumstances that favour it, the suddenness with which it comes about, the wholeness of the conception it embodies, and the fact that the mental events which lead up to it happen below the surface of the mind. But there, to my mind, the resemblance ends. No one questions the inspirational character of musical or poetic invention because the delight and exaltation that go with it somehow communicate themselves to others. Something *travels*: we're carried away. But science is not an art form in this sense; scientific discovery is a private event, and the delight that accompanies it, or the despair of finding it illusory, does not travel. One scientist may get great satisfaction from another's work and admire it deeply; it may give him great intellectual pleasure; but it gives him no sense of participation in the discovery, it does not carry him away, and his appreciation of it does not depend on his being carried away. If it were otherwise the inspirational origin of scientific discovery would never have been in doubt.

[1] See for example J. Bronowski, *Science and Human Values*, revised ed. (New York, 1965).

VICTIMS OF PSYCHIATRY

In a passage in *Bread and Freedom* Camus expresses his revulsion at the way in which, in political arguments, one atrocity may be bartered for another: if one protests at some enormity of the communists, three American negroes are 'thrown in one's face'. In any such disgusting attempt at outbidding, Camus says, one thing does not change – the victim, freedom. It is this reflection that gives Irving Cooper's book[1] its title: the victim is the patient, and the aptness of the Camus quotation becomes very clear as the book goes on.

The subject of Dr Cooper's book is the treatment of a rare and unqualifiedly dreadful neuromuscular disease of stealthy onset known as *Dystonia musculorum deformans* (DMD), in which the limbs and extremities of the victim are locked firmly into grotesquely functionless attitudes by the simultaneous contraction of antagonistic muscles, and in addition victims may suffer mad-looking tremors. It is a disease of the kind that makes even the most devout question the existence of a benevolent deity. Although Lewis Carroll himself thought 'Anglo-Saxon attitudes' a sufficient diagnosis, I put it to Mr Martin Gardner[2] that the White King's looking-glass messenger Haigha, whose extraordinary movements and postures caused Alice so much surprise, was in reality the victim of DMD.

Janet, the patient whose treatment Dr Cooper describes in the greatest detail, contracted the disease at the age of six when she was a reasonably bright girl who weighed about fifty pounds. Five years later, on admission to Dr Cooper's clinic, she weighed only thirty-seven pounds. Her back was arched in such a way as to force her rib cage and stomach forward; the right leg stuck out like a ramrod, with the toes coiled under the sole of the foot, and her left leg was doubled back so that the foot pressed against the buttock. Attempts to move her arms produced clonic contractions, that is, contractions not sustained but compulsively repeated. Clinical examination was made almost

[1] I. S. Cooper, *The Victim is Always the Same* (New York, 1974).
[2] *The Annotated Alice* (Cleveland, 1963).

impossible by the fact that Janet screamed with pain at any attempt to move her limbs and her back arched so much that her head almost touched her buttocks. Her undernourishment was fairly typical and when, as with Janet, the disease follows a chronic and progressive course patients become bedfast and die eventually from inanition or from bedsores.

Although conventional wisdom had long had it that neurosurgery could only substitute paralysis for involuntary muscular contraction Cooper had assembled enough circumstantial evidence to justify the audacious hypothesis that DMD could be meliorated without paralysis by the inactivation through freezing of a group of cells deep in the thalamus, the part of the brain through which sensory impulses pass on their way to the cerebral cortex and which has important motor connections with the cortex.

The operation requires that the head should be kept absolutely still so that a freezing probe, mounted on a stereotactic apparatus, should be guided into the appropriate region of the thalamus. The operation has to be carried out on the conscious patient so that the behaviour of the limbs can give direct evidence of the area of the brain affected (the brain itself does not 'feel pain'). The degree of refrigeration achieved, with liquid nitrogen used as refrigerant ($-196°C$), could be very exactly controlled. The use of refrigeration was a brilliant idea of Dr Cooper's, for it is a procedure less likely than most others to cause a traumatic inflammation which might set up a whole sequence of undesirable side-effects. Cooper is one of the pioneers of cryosurgery and a world-famous master of the use of stereotactic apparatus in the surgery of the brain.

Janet's series of operations had satisfactory results, and most of the book is devoted to her case, but there are three sub-plots: those of David, Donald and Susan. All three stories have it in common that the unhappy patients first fell into the hands of the psychoanalysts – men whose glib and self-assured interpretations of their illness did much to impede diagnosis and treatment. With various refinements all three were judged hysterical. David's affliction was such that when he walked his pelvis was alternately thrust forward and withdrawn, like a caricature of a pop star: clearly exhibitionism, the psychoanalysts opined, and in this case complicated by an extreme neurotic fear of touching his own penis. The psychiatrist's memorandum on David's discharge from the psychiatric ward ended: '*Diagnosis on discharge:* psychoneurosis, conversion hysteria. *Condition on discharge:* Improved.'

After his discharge David got steadily worse and it seems to have been only by coincidence that DMD was eventually diagnosed and treated.

The psychoanalysts' efforts to get at the secret deformities of Janet's psyche included some searching interrogations into her having shared a bath-tub with her brother at the age of seven; she too was labelled a conversion hysteric until a neurologist who saw her later recognised DMD ('there is no question in my mind – a case right out of the book') – equally self-assured, no doubt, but with this important difference: the neurologist was right.

Susan was judged to be writhing and contorting herself in order to attract attention, and her treatment, which so far as it concerned herself was inhumanely cruel, was eventually extended to include her parents – a procedure which nearly broke up their marriage.

It all makes sorry reading, yet Cooper is much more charitable to the psychoanalysts than his readers are likely to be: indeed, he includes a long Fontenelle-like dialogue with Janet in which he explains that it is sometimes genuinely difficult to distinguish psychogenic from physical disorders and that surgeons had sometimes conducted operations to relieve or ascertain the causes of pains of hysterical origin. The overall impression, however, is overwhelmingly condemnatory, and no wonder.

In his wise and temperate foreword Macdonald Critchley, one of the world's leading neurologists, says of DMD:

It is not surprising that the suspicion that the disability is not an organic one can be legitimately entertained for a while. There is no excuse, however, for persisting obstinately in this error, and a note of horror is struck as one reads of the tribulations of Susan and Janet until they eventually received the correct diagnostic appraisal at the hands of neurologists who were belatedly consulted. One likes to think that such a fate could never again befall these victims.

People more affected by manner than by matter will find much to complain of in the style of Cooper's book: he refers to himself throughout in the third person ('his glance rested on the rooftops which obscured Montefiore Hospital'). His style shows more evidence of concern and personal involvement than is traditionally associated with clinical reporting – but why not? I can think of one good reason: it will make Cooper just a little bit more vulnerable to the kind of whispering depreciation that is to be expected from those fashionable psychiatrists (mainly British) who have made such a good thing out of writing imaginative literature on the psyche. Such men use their literary skill not for the advancement of learning but to create a kind of Gothic caricature of psychiatrists who attempt physical remedies: a caricature that typically depicts a man intent upon using powerful psychotropic drugs of unknown action upon a psyche already bruised by electrical convulsions, a man often in collusion with a neuro-surgeon whose lobotomies are remedies hardly more discriminating or sounder in principle than kicking a television set to make it work or stop it flickering.

Alas for them, Cooper is not a good subject for this caricature. His sense of concern and involvement are apparent everywhere and he makes no such general case against them as they will undoubtedly try to make against him. Indeed, the impression the reader will form of Irving Cooper after reading this book is of a skilful and courageous man who in addition to high surgical skill was brilliantly able to take advantage of that felicitous and unpremeditated conjunction of ideas which, following Horace Walpole, has come to be called 'serendipity'.

'Serendipity' is not a good word for Cooper's discoveries (although Macdonald Critchley uses it), because it has too strong a connotation of 'accidental' and not enough of felicity and of the amalgamation of ideas which in a different context would be called wit. However, it should be emphasised that Cooper is a neurologist and physiologist as well as a surgeon and the hypothesis which led him to devise an opera-tion for the relief of DMD is an almost perfect example of what Karl Popper described in a purely methodological sense as a 'risky' hypothesis – the kind of hypothesis that could so very easily be wrong and which for that very reason increases our confidence in its validity when it stands up to empirical trial.

Cooper explicitly says that anybody interpreting his book as a general criticism of psychiatry would be mistaken; but perhaps he is being too generous – for psychoanalysts will continue to perpetrate the most ghastly blunders just so long as they persevere in their impudent and intellectually disabling belief that they enjoy a 'privileged access to the truth'.[1] The opinion is gaining ground that doctrinaire psycho-analytic theory is the most stupendous intellectual confidence trick of the twentieth century: and, to borrow an image I have used elsewhere, a terminal product as well – something akin to a dinosaur or a zeppelin in the history of ideas: a vast structure of radically unsound design and with no posterity.

[1] M. H. Stein, *International Journal of Psychoanalysis*, 53, 13, 1972.

DARWIN'S ILLNESS

CHARLES DARWIN was a sick man for the last forty of his seventy-three years of life. His diaries tell the story of a man deep in the shadow of chronic illness – gnawed at by gastric and intestinal pains, frightened by palpitations, weak and lethargic, often sick and shivery, a bad sleeper, and always an attentive student of his own woes. His complaints began about a year after his return from the great scientific adventures that occupied the five-year voyage of HMS *Beagle*, and they soon took on a fitfully recurrent pattern. Over the next few years, as he became progressively weaker, Darwin gave up his more energetic pursuits, including the geological field-work he had until then delighted in; and in 1842, when only thirty-three, he and his devoted wife Emma retired to a country house in Kent. Darwin left Down House seldom and England never, relying upon correspondence to keep himself up with scientific affairs, and in later years looking fearfully upon the hubbub that broke out after the publication of the *Origin of Species* in 1859.

Like many chronic invalids Darwin came to adopt a settled routine – now a little walk, now a little rest, now a little reading – and three or four hours' work a day was about all he could find energy for. Yet he looked well enough, and was very far from being disagreeable. Every account makes him out considerate and loving, and his granddaughter, Gwen Raverat, described him as affectionate, spontaneous and gay. Nor had he a feeble constitution. He had been an open-air man, strongly built, and at Cambridge a keen shot and sportsman. His records of the *Beagle* and subsequently its Master's show him resilient, tough and full of energy. True, he had not been wholly free of illness. At Valparaiso he had had a fever which, Sir Arthur Keith believed, might well have been typhoid (an important point this, for Keith upheld a psychogenic interpretation of Darwin's later illness); and at Plymouth, waiting fretfully for the *Beagle* to set sail, he complained of palpitations and had dark thoughts of heart disease. But nothing about him gave the slightest premonition of the forty years of invalidism that lay ahead.

What was wrong with Darwin? His own doctors were baffled, and their modern descendants disagree. If Darwin's illness had organic signs they were of a kind his doctors could not then have recognised: they inclined to think him a hypochondriac, and the suspicion that they did so is known to have caused Darwin real distress. Orthodox opinion still has it that Darwin's illness was psychogenic, that is, arose from causes in his own mind; indeed, it figures in Alvarez's textbook on the neuroses as a type specimen of neurasthenia. But what lay behind his neurotic illness? Alvarez, after fifty years' reflection on the matter,[1] seemed to think poor heredity answer enough, and drew attention to the number of difficult and eccentric Darwin and Wedgwood relatives (Emma was a Wedgwood and so was Charles's mother). For Professor Hubble,[2] Darwin's illness, though beyond question of emotional origin, was a subtle adaptation which protected him from the rigours and buffetings of everyday life, the demands of society and the public obligations of a great figure in the world of learning. 'Darwin by his psychoneurosis secretly and passionately nourished his genius' and so gave himself time to execute his great scientific labours: he 'could have done his work in no other way'. There is no refuting Hubble's argument, for there is no argument; the case is presented merely by asseveration ('there can be no doubt', 'it is apparent', 'it is inconceivable', 'it is clear', 'there is overwhelming testimony'). Professor Darlington thinks it possible that the persisting cause of Darwin's illness was the disapproval that grew out of Emma's slow recognition that his doctrines were not such as a Christian might approve of.[3] This, however, is a merely casual suggestion; much more weighty is the full-dress psychoanalytic interpretation of Dr Edward Kempf, one to which The Times felt its readers' attention should be specially drawn.[4]

Kempf believed that Darwin's forty years' disabling illness was a neurotic manifestation of a conflict between his sense of duty towards a rather domineering father and a sexual attachment to his mother, who

[1] New England Medical Journal, 261, 1109, 1959.
[2] Lancet, 244, 129, 1943; 265, 1351, 1953.
[3] C. D. Darlington, Darwin's Place in History (Oxford, 1959).
[4] The Times, 31 December 1963. For Kempf, see Psychoanalytic Review, 5, 151, 1918.

died when he was eight. His mother, a gentle and latterly an ailing creature, fond of flowers and pets, had propounded a riddle which it was Darwin's life-work to resolve: How, by looking inside a flower, might its name be discovered? Kempf wrote in 1918 with an arch delicacy that sometimes obscures his meaning, but Good's[1] more recent interpretation leaves us in no doubt. For Good, 'there is a wealth of evidence that unmistakably points' to the idea that Darwin's illness was 'a distorted expression of the aggression, hate, and resentment felt, at an unconscious level, by Darwin towards his tyrannical father'. These deep and terrible feelings found outward expression in Darwin's touching reverence towards his father and his father's memory, and in his describing his father as the kindest and wisest man he ever knew: clear evidence, if evidence were needed, of how deeply his true inner sentiments had been repressed. 'As in the case of Oedipus, Darwin's punishment for the unconscious parricide was a heavy one – almost forty years of severe and crippling neurotic suffering, which left him at his very best fit for a maximum of three hours' daily work.'

It must be made clear that Darwin's father's tyranny was as unconscious as the hatred it gave rise to. Robert Darwin was a very large (340 lb.) and extremely successful Shrewsbury physician who, starting with the £20 given to him by grandfather Erasmus, made a fortune great enough to support all his children in comfort all their lives. He was a rather overbearing man of decided opinions, and we can see an outcrop of his tyrannical inner nature in his reproaching Charles for his idleness and love of sport at Cambridge and for his getting into the company of what Charles called 'dissipated low-minded young men'. Robert also thoroughly disapproved of Charles's ambition to join the *Beagle*, because he had very much wanted him to go, if not into medicine, then into the Church; but later, at least at a conscious level, he withdrew his objections, and it was he who bought Down House for Charles and Emma.

But much else in Darwin's career must have helped to lay the foundations of a lifetime's neurotic illness. Kempf must, I think, have been the first to call attention – obvious though it now seems – to Darwin's intent and continuous preoccupation with matters to do with sex. We need look no further than the titles of his books: the *Origin* itself, of

[1] *Lancet*, 1, 106, 1954.

course; *Selection in Relation to Sex*; *The Effects of Cross- and Self-Ferti-
lization in the Vegetable Kingdom*; and *On the Various Contrivances by
which Orchids are Fertilized by Insects*. With so great a load of guilt,
need we wonder that at the age of thirty-three Darwin should have
retired from public life to live in quiet seclusion in the country? It was
a sacrificial gesture, even a crucifixion: and Kempf calls attention to
the inner significance of the fact that it was at the age of thirty-three
that Christ himself was crucified.

What is still more important, I feel, is that psychoanalysis has been
able to play a searchlight upon the problem of why Darwin's genius
took its distinctive form. Dr Phyllis Greenacre, in her Freud Anni-
versary Lecture,[1] says she suspects that his turning to science was
mainly the consequence of a 'reaction to sadomasochistic fantasies
concerning his own birth and his mother's death'. But we can be more
particular than this. Kempf reveals to us that when Darwin was specu-
lating upon the selection of favourable variations he was thinking, of
course, of Mother's Favourites ('Darwin was unable to avoid unconsciously founding his sincerest conclusions on his own most delicate
emotional strivings'); and Good explains how in dethroning his
heavenly father Darwin found solace for being unable to slay his
earthly one.

What *was* wrong with Darwin? We may never know for certain,
and there is no other testimony to overwhelm us, but Professor Saul
Adler FRS, of the Hebrew University of Jerusalem, makes a good case[2]
for Darwin's having suffered from a chronic and disabling infectious
illness called *Chagas' disease* after Carlos Chagas Sr, the distinguished
Brazilian medical scientist who first defined it, and caused by a micro-
organism whose name, *Trypanosoma cruzi*, honours another distinguished Brazilian scientist, Osvaldo Cruz. Sir Gavin de Beer, in his
excellent *Charles Darwin*,[3] thinks Adler's interpretation by far the
most likely one. It differs from other theories we have considered in
being based upon the use of reasoning, and Adler's case for it runs
approximately thus:

On 26 March 1835, when spending the night in a village in the
Argentinian province of Mendoza, Darwin was attacked by the huge

[1] *The Quest for the Father* (New York, 1963).   [2] *Nature*, **184**, 1102, 1959.
[3] (London, 1963).

blood-sucking bug *Triatoma infestans*, the benchuca. The benchuca, the 'great black bug of the Pampas', is the chief vector of *T. cruzi*; even today more than 60 per cent of the inhabitants of Mendoza give evidence of the disease and 'as many as 70 per cent of specimens of *Triatoma infestans* are infected with the trypanosome'. It is very likely, then, that Darwin was infected: South American experts consulted by Adler put his chances of escape no higher than 'negligible'. The symptomatology of Darwin's illness can, it appears, be matched closely by known cases of Chagas' disease in its chronic form. De Beer summarises the evidence thus:

The trypanosome invades the muscle of the heart in over 80 per cent of Chagas's disease patients, which makes them very tired; it invades the ganglion cells of Auerbach's nerve plexus in the wall of the intestine, damage to which upsets normal movement and causes great distress; and it invades the auricular-ventricular bundle of the heart which controls the timing of the beats of auricle and ventricle, interference with which may result in heart-block. The lassitude, gastro-intestinal discomfort, and heart trouble from which he suffered an attack in 1873 and died in 1882, all receive a simple and objective explanation if he was massively infected with the trypanosome when he was bitten by the bug on 26 March 1835.

De Beer points out that *T. cruzi* was not identified until twenty-seven years after Darwin's death.

A number of minor and in themselves insubstantial pieces of evidence tell in favour of Adler's interpretation, one of them being that even today inexperienced clinicians may dismiss the chronic form of Chagas' disease as an illness of neurotic origin. I am not aware of any decisive evidence against Adler, and some of the arguments used to discredit his theory establish nothing more than their authors' anxiety to rehabilitate a purely psychogenic interpretation. Disputants so naïve should abstain from public controversy. But clinicians have made it clear to me that the infective theory is by no means a walk-over. There is a general and consistent colouring of hypochondria about Darwin's illness; it is a little surprising that we hear nothing about the acute fever and glandular swellings that would surely have followed infection:[1] and so very long-drawn-out a warfare between host and

---

[1] In this connection, however, my friend Professor P. C. C. Garnham FRS writes: 'The initial infection in many cases is probably unaccompanied by acute

parasite, neither gaining the upper hand for long, is at least unusual. Adler's interpretation has been widely accepted by Brazilian experts, but I suspect that Darwin's having suffered from an illness so closely associated with the names of two great Brazilian scientists is the source of a certain national pride.

The diagnoses of organic illness and of neurosis are not, of course, incompatible. Human beings cannot be straightforwardly ill like cats and mice; almost all chronic illness is surrounded by a penumbra of gloomy imaginings and by worries and fears that may have physical manifestations. I believe that Darwin was organically ill (the case for his having had Chagas' disease is clearly a strong one) but was also the victim of neurosis; and that the neurotic element in his illness may have been caused by the very obscurity of its origins; by his being 'genuinely' ill, that is to say, and having nothing to show for it – surely a great embarrassment to a man whose whole intellectual life was a marshalling and assay of hard evidence. It is a familiar enough story. Ill people suspected of hypochondria or malingering have to pretend to be iller than they really are and may then get taken in by their own deception. They do this to convince others, but Darwin had also to convince himself, for he had no privileged insight into what was wrong with him. The entries in Darwin's notebooks that bear on his health read to me like the writings of a man desperately reassuring himself of the reality of his illness. 'There,' one can imagine his saying, 'I *am* ill, I must be ill; for how otherwise could I feel like this?'

If this interpretation represents any large part of the truth the physicians who inclined to think Darwin a hypochondriac cannot be held blameless, in spite of the fact that the diagnosis of his ailment, if it was indeed Chagas' disease, was entirely beyond their competence. Even among the tough-minded, the mistaken diagnosis of neurotic illness may cause an extreme exasperation – with symptoms which, of course, serve only to confirm the physician in his diagnosis. But Darwin was a

---

symptoms: there is often no more than a small "insect bite" which goes un-noticed amongst all the others, while this inconspicuous lesion is followed by a long period of latency, with no symptoms, until the person falls down dead at the age of 40 or 50 with a ruptured aneurysm; in fact one endemic focus in Brazil is known as the "Land of Sudden Death". Sometimes the course is less dramatic, and, as with Darwin, a chronic illness arises, with signs and symptoms so insidious that the correct diagnosis is often missed.'

gentle creature who had greatly revered his physician father: to such a man the implied diagnosis of hypochondria would carry special authority and do grave and lasting harm. Perhaps Darwin's physicians should have been more on their guard against an interpretation of his illness that gave him so much less comfort than it gave themselves.[1]

[1] See Introduction, p. 5–7, for recent literature on Darwin's illness.

TYPE A BEHAVIOUR AND
YOUR HEART

MORE exigently than any other muscles in the body, the muscles of the heart need an abundant and continuous supply of blood. The blood circulating through the cavities of the heart itself cannot satisfy this need. On the contrary, special arteries – the coronary arteries – supply the musculature of the heart (the 'myocardium') and 5 per cent of the heart's output of arterial blood is appropriated to the coronary circulation. The coronary arteries are thus among the most important in the body, so that coronary arterial disease and coronary heart disease (CHD, which is one of its gravest sequels) are proportionately important as a cause of distress, ill-health or death. Nowadays they affect people in their mid-thirties and early forties as well as in later life. Friedman and Rosenman[1] assemble an amusing list of exhortations or abjurations by which people have sought to keep the horrid enemy at bay. They include: take no animal fats, soft water or sugar, and jog or jog not; even abstention from sexual intercourse is recommended, for it is part of the puritan ethic that any activity so pleasurable must be harmful.

The authors' own message is a simple one: coronary heart disease is of *behavioural* origin and in particular is associated with the behaviour pattern they classify as of Type A:

In the absence of Type A Behaviour Pattern, C.H.D. almost never occurs before seventy years of age, regardless of the fatty foods eaten, the cigarettes smoked, or the lack of exercise. But when this behaviour pattern is present, coronary heart disease can easily erupt in one's thirties or forties. We are convinced that the spread of Type A behaviour explains why death by heart disease, once confined mainly to the elderly, is increasingly common among younger people.

We probably first recognise Type A behaviour in school when a master asks round a class a question to which one boy knows the answer – a

[1] Meyer Friedman and Ray H. Rosenman, *Type A Behaviour and Your Heart* (London, 1974).

boy who waves his arm to and fro in a real agony of superior know-ledge, crying, 'Oh, please sir, ask *me* sir, ask *me*.' Type A behaviour, we are told, refers to a complex of personality traits including:

excessive competitive drive, aggressiveness, impatience, and a harrying sense of time urgency. Individuals displaying this pattern seem to be engaged in a chronic, ceaseless, and often fruitless struggle – with themselves, with others, with circumstances, with time, sometimes with life itself. They also frequently exhibit a free-floating but well-rationalized form of hostility, and almost always a deep-seated insecurity.

The point about the sense of time urgency is well taken: Type As are always listening for radio time-signals and adjusting and readjusting their watches. They also habitually take on more work than they can comfortably do in the time, and time is always an enemy – for 'At their back they always hear Time's wingèd chariot drawing near.' Although Type As are competitive, they seldom compete where they don't excel; the struggle to excel may be deeply exhausting and the failure deeply distressing.

If you hesitate over a word, pretending to stammer, leave a sentence unfinished or stumble over a familiar quotation, someone of Type A is sure to supply the word, to complete the sentence, or put you right. How many readers, I wonder, are quite distressed by their anxiety to inform me that in the couplet from Andrew Marvell quoted above the word 'drawing' should be 'hurrying'?

People of Type A cut such a poor figure throughout this book that I feel something should be said in their favour. Type As are without doubt the great *doers* of the world. Even if Type As lead shorter lives they live much more life while they are living it; it is the existence of people of Type A that makes possible the existence of Type B: do we not all envy those composed and relaxed people who are so adept at getting other people to do their work for them? It is the practitioner of Type A who can always just manage to get one more name on to his list. It is the chairman of Type A who, while slogging away at the re-organisation of his company, makes it possible for his board of Type B directors to congratulate themselves on how well they are doing their jobs. The minor official of Type B is so often the one who couldn't care less: he's far too canny to get into a flap about Mrs

Smith's having nowhere to live because for some accidental reason her name was left off the list of applicants for subsidised housing, and why on earth should he put himself out because some silly young student has lost his passport and with it that golden opportunity to take a lucrative holiday job and learn French at the same time? When it comes to a late-night medical emergency it is canny old Dr B who sagely reflects that if most ailments didn't get better of their own accord nobody would be alive today, and who returns to sleep telling himself that it is part of that anxious young Dr A's training to cope with night emergencies and to learn to distinguish the genuine from the false alarm.

I believe that if a sociologist were to undertake a study of the work habits of professional people, he would find that the habit of systematic overwork was one of the most disagreeable late sequelae of the Second World War, when nearly all non-combatants took the view that over-work was their contribution to the war effort. Nowadays it is still taken for granted that the senior administrator takes home a case full of papers four days a week and should feel guilty if he doesn't have to do so. As for holidays, it is taken as a matter of course that the doctor, scientist, publisher, journalist, engineer and 'executive' find their respective jobs so absorbing that they could perfectly well do without one. This is not because they are severally eaten up by ambition, avarice or lust for self-advancement. It is above all because they have to run faster and faster nowadays in order to remain in the same place (the Red Queen foresaw it all). I suspect that if they were critically analysed, most job specifications these days would turn out to be framed on the assumption of overwork, for a good many responsible jobs are such that they simply cannot be done within the compass of the ordinary working day. It is thus not really fair to imply, as I think these otherwise admirable authors do, in the spirit of Sir William Osler, that the psychosomatic wounds of anxiety and time-harassment are mostly self-inflicted, for they are one of the chief behavioural characteristics of a competitive and somewhat predatory society. 'Just float, Man, just float,' was the advice given to me by a friendly Jesus Freak in North Carolina – ah, but what on? That's the problem. On people of Type A, I surmise.

The authors quote with grim concurrence Sir William Osler's ironic description of how the victims of coronary heart disease are so often the most worthy and exemplary of citizens – up early and late to bed, always striving hard for success in business or professional life, ready at last to take their well-earned ease when their coronary circulations fail. The authors naturally applaud Sir William Osler's critical acumen, for his judgement coincides exactly with their own.

Because they encircle the heart like a crown – almost like a wreath – instead of lying deep in muscle, the coronary arteries are stretched, twisted, turned, jolted with every beat of the heart, not occasionally but over a hundred thousand times a day. 'No flexible tube of any synthetic . . . could withstand this sort of beating for more than a few years, much less a lifetime.'[1]

A good deal of the first part of the book by Drs Friedman and Rosenman is occupied by a description of the anatomy and vascular hydraulics of the heart, a section the reader is told he could skip but advised not to – advice with which I warmly concur.

Friedman and Rosenman interpret the formation of the coronary arterial plaques that narrow the arterial freeway as the outcome of a *reparative* process: tiny wounds of the arterial wall, perhaps caused by the rough usage the coronary artery is subjected to, are healed over by cells derived from endothelium (the innermost lining of all blood vessels), and these little lesions – added to, perhaps, by the proliferation of cells belonging to other layers of the arterial wall – become a focus for the deposition of fatty matter and cholesterol. Well over 50 per cent of American men over the age of twenty-one, we are told, harbour one or more of these plaques. Considering plaques in this light as organs of repair, we can understand that even a large plaque does not in itself threaten life. However, if through impairment of its own blood supply it should come to be replaced by 'necrotic debris' or calcified tissue (coronary 'bone'), then the stage is set for tragedy; for a crack in the plaque itself may now precipitate a clotting process which seriously interferes with the blood supply to the heart and thus causes coronary heart disease, and even the death of heart tissue ('myocardial infarction') – a grave complication of coronary arterial disease.

Elsewhere in the body the slow deprivation of an arterial blood supply can usually be made good by the expansion of a 'collateral'

[1] p. 10.

circulation, that is, the enlargement of the blood supply from an alter-native artery supplying the same tissues. Unhappily, with coronary arterial disease there is no such simple solution, for there are very few interconnecting or bypass channels between right and left coronary circulations. There are, however, a number of tiny connections be-tween the two, and given enough time and demand these enlarge and carry blood from one coronary arterial circulation to the other. Many thousands of people owe their lives to enlargement of these minute connections into functionally adequate vessels. If one asks why nature has been so inept as to deprive mankind of adequate anastomoses between the two coronary circulations, the answer can only be that a disease which has hitherto been chiefly a disease of middle and later life lies largely outside the reach of natural selection: people would have had most if not all of their children before coronary heart disease struck them down, so that the selective penalty was proportionately diminished.

When people ask 'Have the authors proved their case about the relationship between Type A behaviour and CHD?' it seems to me they take a rather simple-minded view of the nature of the scientific process, for theories of this compass and degree of complexity are never 'proved' outright, though sometimes they can be disproved. What happens in practice is that such a theory is received with degrees of approbation or incredulity which vary with each reader's experience and background of theoretical anticipations. From time to time critical efforts are made to find evidence incompatible with the theory, and if these fail and empirical evidence gives any further grounds for having confidence in it then the theory slowly becomes received into the main body of conventional wisdom. I know no evidence that unseats the Friedman–Rosenman theory and it has to my ears a rather frightening authenticity, so I was not at all surprised to learn that William Osler, the great Canadian physician upon the mention of whose name a reverential hush falls over a medical assembly, expressed opinions very similar to those that may be read here.

Even if the causal connection between Type A behaviour and coronary heart disease turns out to be not quite as firm as the authors believe at present, I think their book will still remain an important essay on the conduct of life, for though it is the self-destructive propensities of Type As that the authors naturally concentrate upon, it should not be forgotten that people of Type A can also destroy the composure and ultimately the well-being of people around them – people to whom they do not always contrive to communicate the *hubris* that they may make their own life an exhilarating one.

THE CRAB

ALTHOUGH cancer is much more often curable than its popular reputation leads one to expect, the number of its victims and the sometimes morbid dread (*oncophobia*) of being among them make cancer a source of human distress that is reason enough for the fear it arouses.

In Latin, French, Italian, German and English cancer is *the crab* – so called, perhaps, because of the growth pattern of the most easily visible of all tumours: the rodent ulcer. Some purists (others would say pedants) confine the use of 'cancer' to malignant growths of epithelial tissues, that is, of the tissues that bound surfaces, using 'sarcoma' for malignant growths of connective and supportive tissues and of the white cells of the blood. To laymen these niceties mean nothing. For them a cancer is any threateningly malignant growth.

Dr Thelma Dunn[1] declares that there is an urgent need for an exact definition of cancer that will include all examples of cancer and 'exclude all other abnormal growths or diseases', and she makes much of the admission of Virchow (one of the founders of pathology) that 'no man, even under torture, could say exactly what cancer is'. But in this context, as in so many others, altogether too much fuss is made of matters of definition – for the very ambition to draw a dividing line such as Dr Dunn proposes takes for granted the pre-existence of a working distinction between growths acknowledged to be malignant and others classifiable as benign. My experience as a scientist has taught me that the comfort brought by a satisfying and well-worded definition is only short-lived, because it is certain to need modification and qualification as our experience and understanding increase; it is explanations and descriptions that are needed – and these Dr Dunn provides in abundance.

Hers is not a textbook, though; it could be thought of as a detailed invoice for the expenditure of the very large sums of money allotted to laboratory research – in 1976 a US governmental grant of $396 million

[1] Thelma Brumfield Dunn, *The Unseen Fight Against Cancer* (Charlottesville. 1977).

was spent on basic research, a fourfold increase over the grant in 1970. In its report to the President, the President's Cancer Panel, presided over by Benno C. Schmidt, has already argued most cogently in defence of the expenditure of these very large sums of public money. Dr Dunn's book, which she describes as an informal report, and not for the cancer specialist, might be read as a series of appendices to this report. She writes with the unmistakable air of an insider – a professional who spent the greater part of her working life as a biologist and pathologist at the National Cancer Institute.

Dr Dunn pours scorn on the idea that cancer is essentially a disease of civilisation and the satellite notion that cancer develops in lower animals only after contact with man, but she goes on to explain the demographic circumstances that have nourished this illusion: in most backward societies and in most non-domesticated animals, early death deprives their members of candidature for cancer, the frequency of which tends to increase as life goes on. However, her excellent chapter on cancer-producing agents shows that industrial civilisation cannot be exculpated, for soots, hydrocarbon tars, smokes, a number of food additives, and above all unnaturally high doses of radiation – all unknown in Arcadia – may all cause cancers. Due weight is given to epidemiological research, beginning with Percivall Pott's (1775) classical association of cancer of the scrotum with the occupation of chimney sweep and ending with the identification of tobacco smoke as a principal cause of cancer of the lung – a discovery which has saved thousands of lives and holds the promise of saving thousands more.

More recently still, epidemiological surveys supervised by the Harvard School of Public Health have shown that an early pregnancy confers a certain degree of protection against breast tumours which lasts throughout life. This work will one day make young women reconsider the wisdom of using the Pill to postpone until later on the birth of any child they intend to have.

Dr Dunn allows herself a few laughs at the expense of the practitioners of ostensibly more 'scientific' disciplines such as chemistry and biochemistry: one biologist she refers to was astonished to learn that his (or was it her?) biochemist collaborator supposed that the secretions from the liver and pancreas emptied directly into the stomach. She has some fun, too, at the expense of simple-minded folk who believe that rats have gall-bladders. She concedes, though, that an ignorance almost equally incredible can be found among pathologists. I have myself seen the eyes of a pathologist glaze over momentarily upon my expressing a wish to discuss the at that time widely canvassed possibility that normal and malignant tissues differ in respect of the isoenzymic profiles of their respective lactic dehydrogenases. But does not everyone who uses his mind often find himself stopping miraculously short at the very brink of chasms of ignorance? Test yourself, reader, by engraving upon the head of a pin all you can call instantly to mind about the following subjects: the religious causes of the Thirty Years War, the philosophy of Spinoza, Goedel's Theorem, the nature of deep linguistic structures, and the role of Jan Comenius in the scientific philosophy of the seventeenth century. We can all catch each other out, but fortunately cancer research is a co-operative enterprise in which we all lean upon and sustain each other.

An agreeable feature of Dunn's book is that, so far from regarding all work done before she put pen to paper as of merely antiquarian interest, she introduces many subjects by a brief historical account of how our understanding of them grew. I was particularly pleased to see that the name of Apolant, an assistant of Ehrlich's, had not been forgotten; nor was that of Leslie Foulds. Because of its historical depth, roundedness and intrinsic importance, the chapter on the mammary tumours of mice is perhaps the best in the book and certainly the one that best shows off the author's wide knowledge and sensible judgement.

Dr Dunn refers several times to the formation in 1802 of 'a group of prominent English physicians' to investigate cancer and make authoritative pronouncements upon it. One such pronouncement was 'that no instance had ever occurred or been recorded of cancer being cured by any natural process of the constitution'. Dr Dunn herself adds, 'less than 200 spontaneous cures of cancer are accepted as authentic' – and these are cancers of a special type in the regression of which special factors may intervene.

The acceptance of 200 cases of spontaneous regression in the face of a traditional presumption that no such phenomenon occurs encourages the belief that regressions are very much more common than such a low figure seems to indicate. So far as I am aware, the only agency that might bring about a 'spontaneous' retrogression is an immunological reaction upon the tumour by the organism that bears it. Dr Dunn writes a judiciously and temperately argued chapter on the intervention of immunological factors on the growth of tumours and outlines with complete accuracy the reasons why immunological theories have had their ups and downs of acceptability. I should like to add, however, that in my opinion the epidemiologically established relationship between early pregnancy and the diminution of risk of breast cancer is quite clearly immunological in origin.

Dr Dunn's book comes right into the foreground of modern research. I saw with pleasure (since this is the field of tumour biology in which I work myself) that she cites some of the modern evidence which lends colour to the old-fashioned belief that malignant tumours are to some degree *anaplastic*, that is, re-acquire some of the characteristics of an embryonic cell. Here, perhaps, lies the best hope of the early diagnosis of malignant growths.

One cannot yet attempt a cost–benefit analysis of those hundreds of millions of dollars spent throughout the world on cancer research because not all the credit entries are yet complete. Cancer is not one disease and it will not have one cure, but the way things are going the treatment of a cancer patient is going to acquire more and more the characteristics of a research problem in which the patient, after scrupulously careful biochemical, pathological and immunological assessments, will have a treatment or a system of treatments exactly tailored to suit his condition, by a physician competent to appraise and give due weight to all the evidence that will come before him of the tumour's whereabouts, its degree of malignancy, and the patient's natural power to combat it – which is likely to depend, at least in part, on the soundness of his immunological response system.

It may be that a new kind of physician will be called into being by these demands, much as a new kind of surgeon-scientist was called into being by the growth of therapeutic organ transplantation – a procedure which also makes special demands on the technical skills, knowledge and physiological understanding of the surgeon – a demand which happily was met by the recruitment into surgery of some of the most able students that the medical schools have produced. The sneery attitude toward surgeons that many physicians (particularly in England) were at one time wont to affect is one that nowadays seems most comically ill-judged and, just as many of the leading transplant surgeons today are as much at home in a laboratory as they are in the operating theatre, so it may be that a new generation of oncological physicians will arise who feel just as much at home in the laboratory as in the cancer ward. Just one brilliant break is needed – akin to the first brilliant kidney transplant in the Peter Bent Brigham Hospital in Boston – and then recruits will come forward by the hundred, many of them, I shouldn't wonder, clutching this book by Thelma Dunn in their hot hands.

Unlike Dr Dunn, Larry Agran[1] is an amateur, but he is so intelligent and sensible that it hardly matters. Sometimes, though, in disclaiming personal clinical authority he stands too far back from his subject: 'Cancer, it seems, is not really a single disease susceptible to a single cure.' Why 'it seems'? – surely no one can have tempted him to think that a contrary opinion is widely held.

Agran says that *billions* of dollars have been poured into cancer research in the US and throughout the world, but my guess is that the order of magnitude is 10^8, not 10^9. However that may be, he cites actuarial statistics to show that although some rare tumours (Hodgkin's disease is one) are being mastered, recovery from the commonest and most destructive tumours – mammary and colonic – has not improved as much as this large expenditure authorises one to hope.

What, then, is to be done? asks Agran. An early mention of Percivall Pott (see p. 155), the culture hero of cancer epidemiology, reveals the direction his thoughts are taking, and he goes on to mention the now fully attested cancer-producing actions of radioactive substances and of certain food additives and fibres. He goes on to say that the National Cancer Institute estimates that the vast majority of all human cancers – perhaps up to 90 per cent[2] – are attributable to environmental carcinogens. 'Cancer today, therefore, is actually a disease of man-made origin. What this means, of course, is that *cancer is largely a preventable disease*.'

I think this case is overstated. The idea that cancer is a disease of

---

[1] Larry Agran, *The Cancer Connection: And What We Can Do About It* (Boston, 1977).

[2] Richard Doll, one of the world's foremost epidemiologists, cites a figure of only 80 per cent.

industrial civilisation accords very well, of course, with modern Arcadian thinking, but in a context in which many figures are quoted Agran should surely have attached an exact weighting to his figures of cancer incidence to make clear just how much of the increase is to be attributed to the fact that we live an average of forty or fifty years longer than our ancestors. There can be no cavilling, though, at the proposition that preventable cancers should be prevented.

Where a physician would document his argument with case histories and graphs, Agran does so by skilfully written semi-fictional though factually based narratives relating to victims of environmentally induced cancers. Writing of the activities of OSHA (the Occupational Safety and Health Administration), Agran cites it as one of their axioms that *there is no known "safe" level of exposure to a cancer-causing agent*. The operative word here is 'safe', and although it is by no means impossible that detoxication and repair processes are such that threshold levels of exposure to some carcinogens *do* exist, the cost in time and human lives prohibits any attempt to find out what they are.

I think Agran is a little unworldly in reproaching OSHA for aiming at 'compromise' levels of exposure for political reasons. Surely such compromises are inevitable: we are all sickened by the carnage for which the automobile is responsible, and though it is most doubtful if there is any safe threshold level of speed below which no such carnage would occur, yet we can neither abolish automobiles nor (as in the early days in the United Kingdom) do we require each moving automobile to be preceded by a pedestrian carrying a red flag.

It is, however, not at all unworldly to suggest that manufacturers who expose their workpeople and residents nearby to special hazards should be registered, licensed, required to take preventive measures, and perhaps exposed to unannounced lightning visits by an inspectorate recruited for the purpose. Such considerations apply with special force to the manufacture of asbestos fibres and the more notoriously carcinogenic chemicals such as BCME (bischloromethyl ether – a chemical used in the manufacture of water purification resins).

In the latter part of his book Agran's narrative built around personal case histories is particularly good: some of his stories are moving and disturbing. They may be taken to illustrate how very often it is true to say that the social malefactions blamed upon science and technology or upon 'progress' generally are in reality the outcome of the persistence into the twentieth century of some of the worst elements of nineteenth-century *laissez-faire* capitalism.

The worst sin Agran could be charged with is the use of somewhat heightened colours in the picture he paints; but in his defence it should be pointed out that he is writing of very important matters to do with human welfare, and matters that call urgently for attention. For such purposes as this, monochrome won't do.

Dr Priscilla Laws's book[1] is in the same genre as Agran's and every bit as good. It is about X-radiation. X-rays are good and useful through their power to make the human body, in effect, transparent and thus make possible the recognition or localisation of internal abnormalities. In addition, they cause a disruption of the cell nucleus that is put to good effect in the treatment of cancers, and although X-radiation can sometimes make people very ill, it can have the advantage over drugs taken by injection or by mouth that its action can often be more strictly localised and its dosage more exactly quantified. Above all X-rays can be switched off when they are thought to have done their task.

As against these advantages, X-rays, like gamma rays and other powerful ionising radiations, can themselves cause cancer and – as Herman Muller was the first to point out – can also cause genetic mutation. Dr Laws is not against X-rays in any comprehensive sense but only against the misuse and over-use of diagnostic radiology. Her book is sociologically sophisticated and she sees quite clearly the pressures that have led to the over-use of diagnostic radiology. These include that revolution of rising expectations of health which puts a constant pressure upon the medical profession. It includes the enthusiasm of the radiologist for a skill in which he becomes increasingly proficient as he practises his art, and which makes an important contribution to his standard of living.

Another consideration is less obvious but also important: now that litigation at the expense of the medical profession is becoming something of an indoor pastime in the United States, physicians feel they have to safeguard themselves against much unreasonable litigation. A slight danger of Dr Laws's book is that if laymen and particularly lay women read it uncritically they may be inclined to blame X-radiation for almost

[1] Priscilla Laws, *X-rays: More Harm than Good?* (New York, 1977)

any medical misadventure that befalls them, including miscarriages. Dr Laws is full of anecdotes of which the following is a representative specimen:

'About 15 years ago when I was pregnant, I had a lot of back pain. During my seventh month the pain became more severe, so the doctor took several X rays of my lower back. At the time I was young and very bewildered.

'Well, anyway, the very next day I delivered a 4-pound 11-ounce boy . . . Some years later when my son started school, his teachers told me he was mildly retarded. I don't know if the X rays might have had anything to do with it or not.'

The medical profession is rightly ridiculed by scientists for using anecdotes to bear witness to the therapeutic efficacy of some treatment they are interested in. For fairness' sake it must be said that such evidence is equally inadmissible when it comes to discrediting a medical procedure. Nevertheless the general tendency of Dr Laws's book is good: X-radiation will go on being used and if it is used with more discretion as a result of what Dr Laws has written then she has every reason to feel rather pleased with herself.

Lawrence LeShan's book[1] is about the complicity of emotional and personality factors in susceptibility to and recovery from cancer. 'After two decades of work with cancer patients,' LeShan, a psychotherapist, believes that 'the cancer victim usually has a psychological orientation that increases the chances of getting cancer and makes it more difficult for many individuals to fight for their lives when they do develop a malignancy'.

LeShan's ambitions are very explicitly stated:

First, I intend to set forth the extensive existing evidence that *there is a general type of personality configuration among the majority of cancer patients,* and to suggest some reasons why the emotional responses of these individuals make them more susceptible to cancer.

Secondly, and most importantly, I wish to outline a number of ways in which people *whose personality or life history conforms to this pattern of susceptibility can take steps to protect themselves against the possibility of cancer.*

With these notions very properly in mind (for it is now universally admitted to be an illusion that any research can start with the mind a

[1] Lawrence LeShan, *You Can Fight For Your Life: Emotional Factors in the Causation of Cancer* (New York, 1977).

clean slate) LeShan set himself the task of collecting information about the personality make-up of cancer patients. For this purpose he first applied to them the Rorschach (ink blot) test. He thinks this gave him an insight into the patients' unconscious and made it possible for him to assess the strength and consistency of the ego. But he goes on to explain in psychologese that it didn't actually reveal what they were like, so he dropped the test after only thirty trials and tried next the Thematic Apperception Test (TAT): 'In this test, the patient is shown a series of pictures, of the typical magazine illustration type, and asked to make up a story as to what is happening in the picture.'

TAT was thought to give a broader picture of the patient's personality but 'reactions to it were once again negative'. So eventually the 'Worthington Personal History' test was chosen – something a bit more personal than the conventional curriculum vitae used in applying for jobs. This test, it was felt, 'gave an understanding of the major unconscious stresses, the ego defenses and the techniques of functioning and relating used in everyday life. It also gave a picture of where the patient had been in his life, what he had done and how he felt about different periods of his personal history'.

These various tests, supplemented by carefully thought out personal interviews, occupied the first two years of LeShan's research. But later, when he felt he had gained sufficient basic insight into the particular problems of cancer patients, he 'undertook the more complex and sensitive task of seeing many patients for intensive individual psychotherapy. Over 70 people eventually entered into such treatment.'

On collating his records LeShan found evidence 'in record after record' that the patient had indeed lost the sense of *raison d'être*, but that this had happened some time before the first symptoms of cancer were observed. The patients who had lost their sense of purpose had at one time participated much more fully in life. 'At that time they had had a relationship with a person or group that was of great and deep meaning to them.' On the surface, these people seemed psychologically normal, but underneath there was an absence of direction or goal: 'They felt a lack of any stable reference points for themselves in the universe.'

A second trait LeShan discovered was '*an inability on the part of the individual to express anger or resentment*' – they seeemed to suppress and swallow their hostile feelings, though they did have quite strong feelings of aggression. Third, there were indications that the cancer patient showed some evidences of emotional tension concerning the death of a parent. Not much came of this correlation, though. 'Nevertheless, a start had been made.'

Before turning to an evaluation of LeShan's attempts at psycho-therapy I must comment on the hypotheses as they have unfolded so far. In evaluating an epidemiological theory such as LeShan's, the first stage is to demand clear answers to a range of fairly standard questions: What *exactly* were the controls, that is, in respect of just what characteristics were the controls matched with the *propositi*? Just how many patients with the full gamut of emotional characteristics LeShan refers to did *not* get cancer? And how many contracted ailments *other* than cancer? – a fair question if we have regard to the fact that various personality defects are likely to affect behaviour in ways that diminish the power to combat diseases of other kinds. Again, may not a malignant tumour on the one hand and an emotional disturbance on the other be collateral manifestations of an underlying physical – for example, hormonal – disturbance?

I am sorry to have to say that LeShan's answers to these questions are not such as make it possible to evaluate his theories critically – indeed, his argument has a distressing tendency to sag perceptibly just at points where it ought to tighten up.

I do not put these questions in a hostile spirit, for in spite of several triumphs, it cannot be said that cancer epidemiology is at present so far advanced as to justify a contemptuous or dismissive attitude toward speculations such as LeShan's. Indeed, as an earnest of my antipathy to-wards dismissive criticisms of the possibility of a psychosomatic element in the natural history of cancer I shall cite evidence that may be unfam-iliar to LeShan and that lends unexpected colour to this very possibility. If there is any natural opposition to the growth of tumours, it is almost certain to be immunological in character and to be the outcome of an immunological reaction of much the same kind as that which leads to the rejection of foreign transplants.

Immunity of this kind – 'cell-mediated immunity' – plays an import-ant part in resistance to tuberculosis, an affliction in which a psycho-somatic element is admitted even by those who contemptuously dismiss it in the context of any other ailment. The skin test that is used to reveal a current or previous exposure to tubercle bacilli is known as the Mantoux test or *tuberculin reaction*; a group of workers of the British Medical Research Council led by one of the world's foremost immunologists, Dr J. H. Humphrey, showed conclusively that the outward – in effect vascular – manifestations of the tuberculin reaction could be profoundly influenced by hypnotic suggestion – as neat a demonstration as one could wish of the influence of mind over matter.

In the course of his book LeShan describes his experience of the intensive psychotherapy of seventy patients with 'terminal malignancies'. As his treatments went on he began to get a fuller picture of the character make-up which he thinks causally significant in the genesis of cancer: the patient had a sense of loneliness and unrelatedness and suffered the acquiescent passionless grief of which Elizabeth Barrett Browning has written so movingly. LeShan is quite confident of his ability to distinguish cart from horse and specially emphasises that the world-view which, he says, goes with cancer, 'predates the development of cancer'. This is a risky statement because there may be a long, long gap – perhaps as long as fifty years – between the inception of a tumour and its bursting forth into a malignant growth.

LeShan's clinical experiences are recounted in the form of a number of anecdotes that are interesting and sometimes rather moving. There seems little doubt that he made many of his patients happier as a result of his psychotherapy, but the entire structure of this research is not of a kind that could possibly demonstrate the therapeutic efficacy of psychotherapy in the sense in which, for some tumours, treatment with a drug such as methotrexate is judged to be therapeutically efficacious.

This is all very well, psychotherapists may think, but in real life we can't conduct experiments with the neat cut and dried precision of oncologists working in a backroom with inbred strains of mice. This is a rebuke which all experienced cancer research workers are familiar with and always accept. For all that, the shortcomings of analysis such as LeShan's should not be glossed over: it matters not at all that the patients were in a certain degree self-selected in being well disposed towards the notion of psychotherapeutic treatment and willing to think that it might do them good. What *does* matter is the degree of randomisation in respect to the nature of the cancers treated; for some cancers are notorious for alternating periods of remission with periods of acerbation. Diseases with a natural history such as this – multiple sclerosis is among them – lay cruel traps for would-be therapists: the patients they treat sometimes get better, and sometimes don't – so much so that the people treating them can come to regard themselves as the victims of a conspiracy of nature.

For reasons I hope I have made clear, LeShan's exposition would have benefited greatly if his text had been checked by an all-round tumour biologist before publication. Jane Brody, a medical science writer for the *New York Times*, was wise to write in collaboration with a senior officer of the American Cancer Society.[1] She herself is a bio-chemistry major and a very experienced science writer. The portents are good, then: this ought to be a good book, and I am happy to say that it is. I turned first, out of curiosity, to the pages (271–6) that deal with psychological factors in the causation of cancer, and thought the authors' treatment sensible and temperate. After reviewing various experiments on the relation between stress and cancer, they write:

But all experiments have not yielded consistent results. A psychologist at Kent State University in Ohio found that when rats were subjected to the stresses of overcrowding or electric shock, the development of induced breast cancer in the animals was slowed down and in some cases stopped.

What, if any, significance can be attached to such a collection of findings in animals and man will depend largely on the extent to which scientists are able to decipher the precise biochemical effects of stress and personality patterns. If it can be shown that persons who respond to stress in certain ways suffer meta-bolic, hormonal or immunological upsets that diminish their resistance to cancer, the belief that stress and personality are somehow related to this disease will gain much wider acceptance in the medical community.

As a matter of editorial policy, one very widely read magazine adopts such a rosy-tinted view of all human ailments that one wonders how it comes about that anyone dies at all. Brody and Holleb manage to be simultaneously sanguine and realistic; they hold out hopes where it is reasonable to do so, but are shocking where there is every reason to be so. In addition to the stern lecture on smoking which we expect – and would resent the omission of – there is also some stern writing about

[1] Jane Brody and Arthur I. Holleb, *You Can Fight Cancer and Win* (New York, 1977).

the folly of exposing oneself unduly to the rays of the sun; nor do they gloss over the potential evils of diagnostic X-radiation.

In writing of what can almost be described as an epidemic of diagnostic X-radiography the authors observe that 'while many of these X-rays are important to the patient's welfare and may even be life-saving, according to the United States Bureau of Radiological Health, at least 30 percent of them do not contribute any useful information'. The authors take a conventional but not wholly accurate view of the relevance of immunology to cancer, for research on the subject has already got beyond the level it had achieved when this book went to press. The value of this book to American readers is greatly increased by a series of appendices indicating where a cancer victim can turn for help.

Not long ago I rode on what could have been my last earthly journey, between Rockefeller Hospital and NYU Medical School: the taxi driver, under the impression that I was a practitioner, nearly assaulted me on the grounds that I and my kind had ruined him and his kind because his wife had had a cancer which had ruined him financially, adding (what I did not challenge) that one cannot work more than twenty-four hours in each day. For me the most shocking piece of information in the book by Brody and Holleb was that 'initial treatment of cancer costs an average of $2,000, but it may cost $20,000 or more to care for a patient with advanced cancer'. It is shocking that such statements can be true of any civilised society. With increasing evidence of the contribution of environmental causes to cancer, and the lack of strong fiscal or commercial incentives to correcting them, a society can no longer regard the treatment of cancer as a personal transaction between the patient and his physician, particularly if a patient's whole family can be ruined by a disease which he may have contracted through no fault of his own. Federal donations to Cancer Research are in reality not much more than conscience money: nothing will do except legislation on a national scale that will save a patient's livelihood no less effectively than his life.

UNNATURAL SCIENCE

IF a broad line of demarcation is drawn between the natural sciences and what can only be described as the unnatural sciences, it will at once be recognised as a distinguishing mark of the latter that their practitioners try most painstakingly to imitate what they believe – quite wrongly, alas for them – to be the distinctive manners and observances of the natural sciences. Among these are:

(*a*) the belief that measurement and numeration are intrinsically praiseworthy activities (the worship, indeed, of what Ernst Gombrich calls *idola quantitatis*);

(*b*) the whole discredited farrago of inductivism – especially the belief that facts are prior to ideas and that a sufficiently voluminous compilation of facts can be processed by a calculus of discovery in such a way as to yield general principles and natural-seeming laws;

(*c*) another distinguishing mark of unnatural scientists is their faith in the efficacy of statistical formulae, particularly when processed by a computer – the use of which is in itself interpreted as a mark of scientific manhood. There is no need to cause offence by specifying the unnatural sciences, for their practitioners will recognise themselves easily: the shoe belongs where it fits.

The objections of the educated to IQ psychology arise from several sets of causes: first, misgivings about whether it is indeed possible to attach a single-number valuation to an endowment as complex and as various as intelligence; second, a biologically well-founded feeling of repugnance to the notion that differences of intelligence are to so high a degree under genetic control that all the apparatus of pedagogy and special training is necessarily relegated to an altogether minor role. To these have recently been added a third, some grave doubts about the probity of Cyril Burt's investigations of intelligence quotients in twins – researches which led him to conclusions which have had a profound and by no means wholly beneficent effect on educational theory and practice. Burt's work has been the subject of extensive correspondence and annotation in both the London *Times* and the *Sunday Times*.

We must consider first the illusion embodied in the ambition to attach a single-number valuation to complex quantities – a problem that has vexed demographers in the past, and also soil physicists – as Dr J. R. Philip, FRS, has pointed out.[1] It bothers economists, too.

Although the more disputatious IQ psychologists give the impression of being incapable of learning anything from anybody, it seems only fair to give them a chance not to persist in the errors of judgement that have been avoided in so many other areas of learning. Let us discuss the single-number valuation of complex variables in a number of different contexts.

First, demography. In the days when it was believed that the people of the Western world were dying out through infertility, it was thought an obligation upon demographers to devise a single-value measure of a nation's reproductive prowess and future population prospects. Kuczynski accordingly offered up his 'net reproduction rate' and R. A. Fisher and A. J. Lotka the 'Malthusian parameter' or 'true rate of natural increase'. Both had their adherents, and confident predictions were based on both, but the predictions were mistaken and today no serious demographer believes that a single-number valuation of reproductive vitality is feasible: reproductive vitality depends on altogether too many variables, not all of which are 'scalar' in character. Among them are the proportions of married and of unmarried mothers, the prevailing fashions relating to marriage ages, family numbers and the pattern of family building, the prevailing economic and fiscal incentives or disincentives to procreation, and the availability and social acceptability of methods of birth control. It is no wonder that the single-number valuations of reproductive vitality have fallen out of use. Modern demographers now go about their population projections in a biologically much more realistic way, basing them essentially upon the sizes of completed families and the analysis of 'cohorts' – groups of people born or married in one specific year.

Somewhat similar considerations apply to the attempt to epitomise in a single figure the field behaviour of a soil. The physical properties and field behaviour of soil depend upon particle size and shape,

[1] 'Fifty Years' Progression in Soil Physics', *Geoderma*, **12**, 265–80, 1974.

porosity, hydrogen ion concentration, bacterial flora, and water content and hygroscopy. No single figure can embody in itself a constellation of values of all these variables in any single real instance.

Rather similar considerations apply to the way some economists use the notion of GNP ('the tribal God of the Western world'). GNP as such may be an unexceptionable idea, but there has been an increasing tendency to use the growth rate of GNP, positive or negative, as a measure of national welfare, well-being, and almost of moral stature. Any such use is, of course, totally inadmissible: how can a single figure embody in itself a valuation of a nation's confidence in itself, its practical concern for the welfare of its citizens, the stability of its institutions, the safety of its streets, and other such non-scalar and therefore presumably unscientific variables?

IQ psychologists would nevertheless like us to believe that such considerations as these do not apply to them; they like to think that intelligence can be measured as if it were indeed a simple scalar quantity. I recall in particular the barefaced impudence with which a notorious IQ psychologist has proposed that a person's IQ *is* his intelligence as much as his height might be 5 feet 5 inches. Unhappily for IQ psychologists, this is not so. If they were merely playing an academic game that did not affect the rest of us for good or ill, they would of course be entitled to define intelligence in any way they wished, but for the educated, 'strength of understanding', as Jane Austen described it, is a complicated and many-sided business. Among its elements are speed and span of *grasp*, the ability to see implications and conversely to discern a *non sequitur* and other fallacies, to discern analogies and formal parallels between outwardly dissimilar phenomena or thought structures, and much else besides. One number will not do for all these, even if – to take what must surely be one of the most abject of arguments put forward by IQ psychologists in favour of single-value mensuration – a child's IQ score is positively correlated with his income in later years.

To turn now to the vexed problem of the heritability of intellectual differences, it may be said with some confidence that unless intellectual abilities are unlike all others and unless human beings are unlike all other animals in respect to possessing them – two suppositions that are by no means as far-fetched as we may at first incline to think them (see below) – then intellectual differences are indeed genetically influenced. This applies even if upbringing and indoctrination are of preponderant importance: for here we should certainly expect inherited differences in teachability and the ability to profit by experience.

The subject is bedevilled more than any other by the tendency of disputants to spring into political postures which allow them no freedom of movement. Thus it is a canon of high Tory philosophy that a man's *breeding* – his genetic make-up – determines absolutely his abilities, his destiny and his deserts; and it is no less characteristic of Marxism that, men being born equal, a man is what his environment and his upbringing make of him. The former belief lies at the root of racism, Fascism, and all other attempts to 'make Nature herself an accomplice in the crime of political inequality' (Condorcet), and the latter founders on the fallacy of human genetic equality ('A strange belief', said J. B. S. Haldane – though a long-time member of the CP).

Confronted with this dilemma, modern liberals are keenly aware that, not so very long ago, there were countries in which those who questioned the dogma of genetic élitism would have been trampled down by big boots; but they have been slow – as liberals sometimes are – to realise that today it is the other way about; those whose views conflict with the dogma of equality are vilified, shouted down and rebutted by calumnies. Human geneticists are particularly vulnerable to the vilification of doctrinaire Marxists because, as scientists, they are in thrall to such bourgeois superstitions as the desirability of telling the truth. Among the latest victims of such vilification are the human geneticists engaged in human karyotype screening, which entails the investigation of the human chromosome make-up at birth or earlier, to identify in good time such abnormalities as are now known to be associated with Down's syndrome ('mongolism'), a number of disorders of sexual development (for example, Turner's syndrome, Klinefelter's syndrome), and sometimes grave personality disorders, particularly that which is associated with the human sex chromosome make-up symbolised as 47XYY. The president of the American Society of Human Genetics, Dr John L. Hamerton, delivered a wise and temperate address on the problems raised by karyotype screening at the annual meeting of the society in Baltimore in 1975.[1]

[1] *American Journal of Human Genetics*, **28**, 107–22, 1976.

Chromosomal abnormalities are unfortunately irremediable, but this is not to say that, with advance warning, their physical and behavioural consequences cannot be the subject of meliorative or preventive intervention. Nevertheless, malevolent intentions are taken for granted by disputants claiming to speak – as they all do – for 'the people'.

The really important question, however, is whether or not it is possible to attach exact percentage figures to the contributions of nature and nurture (Shakespeare's terminology) to differences of intellectual capacity. In my opinion it is *not* possible to do so, for reasons that seem to be beyond the comprehension of IQ psychologists, though they were made clear enough by J. B. S. Haldane and Lancelot Hogben on more than one occasion, and have been made clear since by a number of the world's foremost geneticists.

The reason, which *is*, admittedly, a difficult one to grasp, is that the contribution of nature is a function of nurture and of nurture a function of nature, the one varying in dependence on the other, so that a statement that might be true in one context of environment and upbringing would not necessarily be true in another. To choose an extreme example: the low-grade mental deficiency known to be associated with a constitutional inability to handle the dietary ingredient phenylalanine is a departure from normality that might be judged simply hereditary in children brought up on a normal diet abundant in the dietary constituent they cannot handle; for phenylketonuria is certainly due to a conjunction of genes that are inherited according to straightforward Mendelian rules.

If, however, a newborn child with the make-up that would otherwise have made it a victim of phenylketonuria is brought up in a microcosm free from phenylalanine – a difficult and expensive feat – then phenylketonuria would not make itself apparent. In this extreme case therefore a situation can be envisaged in which the disability is wholly environmental in origin. It will manifest itself in the presence of phenylalanine but not in its absence, and will thus present itself as a disease caused by phenylalanine. Alternatively, in a real world abundant in phenylalanine we can confidently describe the departure from normality as genetic in origin.

This example is perhaps too extreme to be informative, so I shall use instead an example which may help to make the point more clearly. The little brackish water shrimp *Gammarus chevreuxi* is extruded from the brood pouch with red eyes, but usually ends up with black eyes – because of the deposition in them of the black colouring matter melanin. The capacity for forming melanin and the rate at which it is formed and deposited are between them under the control of a number of genetic factors. Colouration of the eye is also affected by a number of other environmental factors: certainly the temperature and probably (though I don't know for sure) the dietary availability of such substances as tyrosine and phenylalanine or their precursors.

Among these various factors temperature is perhaps the most instructive, for it is possible to choose a genetic make-up such that colouration of the eye will appear to be wholly under environmental control: black at relatively high temperatures of development and reddish or dusky at lower temperatures. It is also possible to choose an ambient temperature at which red eyes or black eyes are inherited as straightforward alternatives according to Mendel's laws of heredity. Thus to make any pronouncement about the determination of eye colour it is necessary to specify both the genetic make-up and the conditions of upbringing: neither alone will do, for the effect of one is a function of the effect of the other. It would therefore make no kind of sense to ask what percentage of the colouration of the eye was due to heredity and what percentage was due to environment.

In an earlier paragraph I referred to the extreme likelihood of heredity's playing some part in the determination of differences of intellectual performance, adding, however, for form's sake, the qualification 'unless intellectual abilities are unlike all others and unless human beings are unlike all other animals in respect to possessing them'. This possibility I should now like to consider in the light of modern ethological research and our newer philosophic understanding of the character of cultural inheritance in mankind.

Human beings owe their biological supremacy to the possession of a form of inheritance quite unlike that of other animals: exogenetic or exosomatic heredity. In this form of heredity information is transmitted from one generation to the next through non-genetic channels – by word of mouth, by example, and by other forms of indoctrination; in general, by the entire apparatus of culture. I have illustrated this idea[1] by pointing out that it was not the making of a wheel that represented a characteristically human activity, but rather the communication from one person to another and therefore from one generation to the next of the know-how to make a wheel. In this view, Man is not so much a tool-making as a communicating animal. Exogenetic or cultural heredity is that which has made possible the inauguration and retention of the cultural and institutional elements of our current civilisation.

Apart from being mediated through non-genetic channels, cultural inheritance is categorically distinguished from biological inheritance by being Lamarckian in character; that is to say, by the fact that what is learned in one generation may become part of the inheritance of the next. This differentiates our characteristically human heredity absolutely from ordinary biological heredity, in which no specific instruction can be imprinted upon the genome in such a way as to become part of the package of inheritance: in ordinary evolution genetic processes are *selective* and not *instructive* in character: genetic changes do not arise in response to an organism's needs and do not, except by accident, gratify them. There is no great mystery about what has made this new pattern of heredity and evolution possible: it has been made possible by the evolution of an organ, the brain, of which the main function is to receive information from the environment and to propagate it. In such a system of heredity, indoctrination on the one hand and on the other hand imitation ('aping') and teachability play crucially important parts – as they are already known to do in the behaviour of cats and of apes.

It is very likely therefore that selective forces acting on mankind will have promoted the power of the brain to receive and communicate information, and will have made teachability an endowment of premier importance, so that, while there are likely to be inherited differences of teachability, it is extremely unlikely that teaching and training cannot improve intellectual performances. Indeed, if an intellectual performance were to be totally unaffected by training and practice I should be inclined to think that the wrong performance was being measured. It is because of the embarrassingly foolish belief that an IQ performance measures a person's 'innate intelligence' that

[1] 'Technology and Evolution', pp. 184–90 below.

extreme hereditarians take the view that IQ is invariant under educative procedures — a claim that reminds one of Francis Galton's contempt for those who try to raise themselves beyond whatever station in life it may have pleased their genes to call them to.[1] If it were indeed true that IQ is invariant with age then the only conclusion we could legitimately come to is that the tests upon which its measurement is based are tests of the wrong capacities.

In short, although the possibility of its being so was introduced more as a formal disclaimer than with any other serious purpose, we can conclude that the pattern of inheritance of intellectual differences in human beings is indeed different from the inheritance of other character differences in other animals.

In his *The Science and Politics of IQ*,[2] Leon J. Kamin is 'concerned with a single major question: are scores on intelligence tests (IQ's) hereditable?'. The answer, he says,

in the consensus view of most intelligence testers, is that about eighty per cent of individual variation in IQ scores is genetically determined. This is not a new conclusion. Pearson, writing in 1906, before the widespread use of the IQ test, observed that 'the influence of environment is nowhere more than one-fifth that of heredity, and quite possibly not one-tenth of it.' Herrnstein, reviewing the history of intelligence testing to 1971, concluded 'We may therefore say that 80 to 85 per cent of the variation in IQ amongst whites is due to the genes.'

Kamin goes on to state it as a principal conclusion of his book that: 'There exist no data which should lead a prudent man to accept the hypothesis that IQ test scores are in any degree heritable', and then asks how it is that a contrary opinion has so long prevailed among psychologists. Kamin himself believes that socio-political motives underlie the willing assent of IQ psychologists to the notion of inherited differences in intellectual capacities. Indeed, he carries this conspiracy theory of heritability to the point of suggesting that the entire project of IQ psychology is implicitly a great salve for the public conscience and incidentally a great relief to the public purse: if the poor are unsuccessful and inferior because they have been born that way and not because of the way they have been treated, then there is not much we can do about it.

[1] 'Taking the Measure of Man', pp. 203–8 below.
[2] (Potomac, Md, 1974).

Thus the extreme hereditarian viewpoint is seen as part of that great conspiracy referred to above to 'make Nature herself an accomplice in the crime of political inequality'. The conspiracy is not, of course, declared and open, but is rather the subconscious consequence of these economic and class-competitive forces that are thought to shape history. Thus Kamin's interpretation of the origins of hereditarian theory has about it the kind of Olympian glibness more often found in psychoanalytic theory, and it is equally difficult to refute. For just as any criticism of psychoanalysis is construed as an infirmity of the psyche which itself requires psychoanalytic treatment, so criticism of an essentially Marxist theory is thought to reveal its author as yet another victim and dupe of the very socio-economic forces whose efficacy he has presumed to question.

In writing of the pioneers of IQ testing, Kamin makes the useful point, quite new to me, that when Alfred Binet pioneered intelligence testing he described as 'brutal pessimism' the belief that the intellectual performance of an individual could not be augmented by special training, and indeed prescribed a therapeutic course in 'mental ortho-paedics' for those with lowly test scores.

Binet was an agent of the State schooling system in France, and the purpose of his intelligence tests was to identify children in need of special schooling. It was far otherwise with Binet's American heirs, particularly Lewis M. Terman, who came to regard an intelligence test score as a measure of a fixed quality thought of as 'innate intelligence' – an expression still in use and as clearly indicative today as it ever was of a deep-seated misunderstanding of genetics. Moreover, a political, racist and – in the worst sense of the word – eugenic motivation is made painfully clear by some of Kamin's quotations from the pioneers of IQ testing. They may not have been the worst offenders: they were writing at a time when it was widely believed that the riotous proliferation of the feeble-minded would repopulate the world with imbeciles, and that affections such as 'mongolism' (Down's syndrome) represented an atavistic degeneration to a primitive and lowly human type (hence the name).

Nevertheless, the alleged malevolence, racial bias or even down-right dishonesty (see below) of hereditarian psychologists cannot answer the material question whether or not heredity contributes anything to differences of intellectual performance. In denying any such influence, Kamin goes too far – just as H. J. Eysenck went too far in a passage the mere contemplation of which probably now causes him acute embarrassment: 'the whole course of development of a child's intellectual capabilities is largely laid down genetically'.

With thinkers such as Terman to guide them, we need not wonder at how nearly castration became a statutory requirement in a number of American States, nor at how confidently one State legislature or another resolved that heredity played a principal part in crime, idiocy, imbecility, epilepsy and dependence on charity.

The first tests to reveal that Blacks score less than Whites emerged from the extensive screening undertaken in the First World War, tests of which Kamin drily remarks that they 'appear to have had little practical effect on the outcome of the war'. Such tests have, however, had a profound effect on the relationship between Blacks and Whites ever since. Another important part of the harvest of the routine screening of recruits was a vast heap of unreliable information on the intelligences of recruits classified by their countries of origin – evidence from which it became pretty clear that northern European countries scored highest, with Mediterraneans, Slavs and other such lowly types a good way behind. These findings became known in Congress and had an important influence in shaping the US immigration laws.

Madison Grant, in his *The Passing of the Great Race*, lamented the likelihood that the American people would be irretrievably diminished by the influx of inferior foreigners. With his nice touch for allowing the subjects of his criticism to assassinate themselves, Kamin quotes passages from Grant and a Professor C. C. Brigham of Princeton that sound like a crash course in racism:

The Nordics are . . . rulers, organizers, and aristocrats . . . individualistic, self-reliant, and jealous of their personal freedom . . . as a result they are usually Protestants . . . The Alpine race is always and everywhere a race of peasants . . . The Alpine is the perfect slave, the ideal serf . . . the unstable temperament and the lack of coordinating and reasoning power so often found among the Irish . . . We have no separate intelligence distributions for the Jews . . . Our army sample of immigrants from Russia is at least one half Jewish . . . Our figures, then, would rather tend to disprove the popular belief that the Jew is intelligent

. . . he has the head form, stature, and color of his Slavic neighbors. He is an Alpine Slav.

It seems to me that many of the socially disruptive influences that have been drawn from the study of IQ performances are the consequence not so much of the malevolence of those who undertake them as of the inherent failings of IQ testing itself. The illusion that a single-number valuation can be attached to anything that an educated man means by the word 'intelligence' has already been exposed. But a still graver illusion is even more dangerous because it places foreigners, the poor and the deprived at a special disadvantage – the illusion that intelligence tests can be devised which are 'culture free', that is, which are quite uninfluenced by the cultural background of the subject's family, or by the linguistic or performative exercises which he may, or more likely may not, have taken before testing. These naïve beliefs are now passing out of favour, but not before they have done a very great deal of harm.

Where so many hereditarian writers are graceless, rancorous and inept, Kamin writes with a winning skill that Jonathan Swift would have delighted in. He is merciless to the Californian sickness:

The meek might inherit the kingdom of Heaven, but, if the views of the mental testers predominated, the orphans and tramps and paupers were to inherit no part of California. The California law of 1918 provided that compulsory sterilizations must be approved by a board including 'a clinical psychologist holding a degree of Ph.D.'. This was eloquent testimony to Professor Terman's influence in his home state.

These passages of fine polemical writing must not be allowed to distract attention from the most important part of Kamin's book, his critique of observations purporting to demonstrate a very high degree of heritability of differences in IQ scores. Because it has been the subject of some searching investigative journalism by the *Sunday Times* in London, we shall pay particular attention to the testing of twins.

Kamin gives an admirably lucid account of the methodology of twin studies, of which the underlying principle is this: twins may be of the kind called identical, that is, the product of a single fertilised egg, or they may be 'fraternal', that is, litter mates – who resemble each other genetically no more closely than ordinary brothers and sisters. Identical twins can be assumed fairly confidently to have the same genetic make-up. Identical twins who have been separated and brought up in different environments are methodologically a godsend. The degree of correlation between their measured intellectual performances is an estimate of the degree to which heredity has contributed to them, *provided* the various environments are representative of the whole range of environments to which human beings are exposed, and twins themselves are representative of the entire population of which they are members.

However, as Kamin writes, 'there is little reason to suppose that these assumptions hold in any of the studies that have been made of separated twins'. Kamin pays special attention to the studies made by Professor Cyril Burt, one of the great pioneers of educational psychology, and Professor of Psychology at one of England's three leading universities: University College, London. Burt's direct influence was probably, largely, a harmful one; because of his advocacy and the tendency to regard his opinions as Holy Writ, eleven-year-olds in Great Britain were subjected to tests intended to divide the bright from the comparatively dull. Indirectly his teachings may be said to have invited the backlash which has led now to the reinstitution of those comprehensive schools that are founded on the proposition that all children are fundamentally of equal ability – so making the usual confusion between the fact of biological inequality and the political right to equal treatment. Still, he can hardly be blamed if, for political reasons, his teachings have now had the effect of handicapping those very children whose interests they were designed to promote.

Kamin's criticisms of Burt make some of the most damaging accusations that can ever have been levelled against a scholar:

the various papers published by Burt often contain mutually contradictory data, purportedly derived from the same study. These contradictions, however, are more than compensated for by some remarkable consistencies which occur repeatedly in his published works. The first examples that we shall cite do not involve his study of separated twins, but later examples will do so.

The papers of Professor Burt, it must be reported, are often remarkably lacking in precise descriptions of the procedures and methods that he employed in his IQ testing. The first major summary of his kinship studies, a 1943 paper, presents a large number of IQ correlations, but virtually nothing is said of when or to whom tests were administered, or of what tests were employed. The reader is told, 'Some of the inquiries have been published in LCC reports or elsewhere; but the majority remain buried in typed memoranda or degree theses.'

Toward the end of 1976 a furore was started by the publication in the London *Sunday Times* of an article by a team of investigative journalists led by Dr Oliver J. Gillie, their medical correspondent and a gifted geneticist. The investigations questioned the probity of Burt's entire work, raising a number of awkward questions to which no satisfactory answers had then been given. In addition, Professor Jack Tizard, the highly respected Professor of Child Development in London University, delivered a lecture likening the revelations about Burt to those which disclosed the forgery of the Piltdown skull. Tizard said his suspicions had been aroused two and a half years beforehand by his complete failure to find two people at University College who were said to have worked very closely with Burt in his research – colleagues with whom he had, indeed, published a number of papers between 1952 and 1959 – namely Miss Howard and Miss Conway.

The *Sunday Times* team fared no better; they could find no record that either had ever been on the staff of the Psychology Department at the University College, and could not even trace them in the files at Senate House, reputedly the central nervous system of the University of London, which holds duplicates of the documents of the University's constitutent colleges: 'Direct inquiries to 18 people who knew Burt and his circle well from the 1920s, when he was at the National Institute of Industrial Psychology, until he died, have failed to find anyone who met Howard or Conway or knew of them, and no one with these names is listed in the files of the British Psychological Society.' The *Sunday Times* concluded its investigation by proposing the hypothesis that Misses Howard and Conway never existed.

However, in spite of these misgivings there seems no doubt that Margaret Howard anyway did 'in some real sense' (as philosophers say) exist. Professor John Cohen of Manchester University is quoted by Dr Gillie as saying that he knew Miss Howard well. In the follow-up article containing this revelation he also quotes a damaging accusation by Professor and Mrs Clarke of Hull University that articles which they did not write were published in their name by Sir Cyril

Burt, and they add, 'It is extremely difficult to see how Burt arrived at some of his figures on inheritance of intelligence without cooking them.'

Kamin's evidence and the *Sunday Times* investigations point to Burt's having a fairly lofty attitude toward the provenance and probity of his empirical data. Indeed, the accusation that Burt's findings were too good to be true – that is, were too closely in line with theoretical anticipations – gives us a clue to the most puzzling question of all about Burt: Why, why did he act deviously?

The only explanation I can think of is that a belief in the predominant influence of heredity in relation to intellectual performances has the same kind of appeal for those who hold it as Lamarckism – the belief in an inherence of acquired characters – has had for unskilled biologists. It seems to them so *right*, so obviously and necessarily true, so clearly in keeping with their sense of the fitness of things, that people who do not share their beliefs must somehow be persuaded in their own best interests to do so, if necessary by a slight adjustment of the figures here, an assumption based upon a lifetime of experience there, and judicious selection of data somewhere else. Fraudulent experiments have been used to uphold Larmackian interpretations of heredity, and in Burt's methodological malpractices we may have another case in point. Villainy is not explanation enough: Burt probably thought of himself as the evangelist of a Great New Truth.

So much anyway for the case for the prosecution. The most significant utterance in the case for the defence is that of Professor Eysenck, himself a dedicated hereditarian. In a letter in the *Sunday Times* citing Burt's data and calculations he concedes that some of his procedures were 'of course inadmissible' to a degree 'that makes it impossible to rely on these figures in the future'.[1] On the other hand Professor A. R. Jensen, joining in the *Times* debate, did not share Eysenck's view that any of Burt's procedures were inadmissible: he dismisses the attack on Burt as so much calumny and concludes, with the 'complete confidence' which natural scientists so seldom feel, that 'even if all of Burt's findings were thrown out entirely, the picture regarding the

---

[1] Eysenck developed his interpretation in *Encounter* for January 1977 and says that the most Gillie was entitled to say was that 'there were certain inconsistencies in Burt's data which called in question the interpretation which might be put upon them'.

UNNATURAL SCIENCE 181

heritability of IQ would not be materially changed'. I am quite sure Jensen is not intending to be ironical; but this judgement does seem to be a rather strange compliment to a man thought of as a founding father of psychometry.

There is, as a matter of fact, a well-established precedent for the selection or adjustment of figures to fit a preconceived hypothesis: R. A. Fisher, at that time the world's foremost authority on small-sample statistics, once pointed out that Mendel's famous segregation ratios (3:1; 9:3:3:1) were numerically much too good to be true. Given the size of his samples, no such degree of conformity to theoretical anticipation could be judged plausible. Whatever R. A. Fisher's motives may have been in calling attention to this fact, we may be quite sure it was not his intention to show Mendel up as a running-dog of Fascism (as the faithful later came to call him). The most plausible explanation seems to be that the abbé's gardeners and assistants had formed a pretty clear idea of what ratio Mendel was expecting, and whether out of loyalty or affection supplied their reverend employer with results they thought he would like to hear.

There is, however, a profoundly important difference between the cases of Mendel and of Burt: Mendel was right.

Now that the IQ controversy has risen to a new height and shows no sign of abating, the publication of Block's and Dworkin's *The IQ Controversy*[1] is particularly timely and valuable. It is in the genre known as a 'reader', that is to say it gives us a conspectus of prevailing opinions in the words of those who hold them. The danger of a reader in such a context as this is that the editor may, by judicious selection or omission, prejudice the conclusions that an impartial reader might come to. Block and Dworkin have not done this: their editorial matter provides only that minimum of connective tissue which a book such as this urgently needs. It is also fair to point out that the only way in which the hereditarians could be rescued completely from public obloquy would be by omitting their contributions altogether. Block and Dworkin have rightly decided against so partial a procedure: they play fair, though it might be thought cruel to republish the controversy between Walter Lippmann and Lewis Terman, published in the

[1] Ned Block and Gerald Dworkin (eds), *The IQ Controversy: Critical Readings* (New York, 1976; London, 1977).

New Republic in 1922 and 1923. This gives one the sick feeling that people of sensibility have when they first witness a bullfight: the contest is so cruelly unequal when one contestant has nothing but a slow-footed and ponderous irony with which to defend himself against the highly intelligent, light-footed and cruelly provocative Mr Lippmann.

A special strength of their book – and one that enormously enhances its value for college reading – is their generous allocation of space to such real professionals as Richard Lewontin and John Thoday, with a passing quotation from Michael Lerner. In the course of a grave, learned and witty investigation of what has come to be called 'Jensenism', Lewontin remarks: 'There is no such thing as *the* heritability of IQ, since heritability of a trait is different in different populations at different times. Second, the data on which the estimate of 80 per cent for Caucasian populations is based, are themselves of very doubtful status.'

The citation from Michael Lerner includes this sentence: 'it is a fact that generations of discrimination have made direct comparisons of mental traits between Negroes and whites not biologically meaningful'.

John Thoday expounds clearly and critically the methodology of intra-group and between-group comparisons, calling attention, as he does so, to blunders by IQ psychologists of a kind that disclose a truly deep-seated misunderstanding of genetic principles. He concludes that 'there is no evidence which reveals whether the Negro–white IQ difference has any genetic component or any environmental component'.

The reflection that might well be in the forefront of the minds of laymen as they put down the Block and Dworkin book is this: The question of the heritability of differences of IQ is one upon which everybody feels entitled to have an opinion. In recent years even a prominent electrician has felt authoritative enough to have his say; yet on matters to do with heritability it might be thought prudent to give most weight to the opinions of geneticists. Why, then, is it that some of the world's most prominent geneticists – among them Michael Lerner, Richard Lewontin, Walter Bodmer and John Thoday – remain so deeply unconvinced by the hereditarian arguments of such as Jensen and Eysenck? We need not resort to murky ideological explanations to find the reason. It is more likely, I suggest, that at a time of deeply troubled race relations, when the whole possibility of peaceful co-existence and mutual respect in multiracial communities is in question in many parts of the world, these geneticists feel an imperatively urgent desire to put the scientific record straight.

TECHNOLOGY AND EVOLUTION

Genetic and exogenetic heredity

The use of tools has often been regarded as the defining characteristic of *Homo sapiens*, that is, as a taxonomically distinctive characteristic of the species. But, in the light of abundant and increasing evidence that subhuman primates and even lower animals can use tools, the view is now gaining ground that what is characteristic of human beings is not so much the devising of tools as the communication from one human being to another of the know-how to make them. It was not so much the devising of a wheel that was distinctively human, we may suggest, as the communication to others, particularly in the succeeding generation, of the know-how to make a wheel. This act of communication, however rudimentary it may have been – even if it only took the form of a rudely explanatory gesture signifying 'It's like this, see', accompanied by a rotatory motion of the arm – marks the beginning of technology, or of the science of engineering.

Everyone has observed with more or less wonderment that the tools and instruments devised by human beings undergo an evolution themselves that is strangely analogous to ordinary organic evolution, almost as if these artefacts propagated themselves as animals do. Aircraft began as birdlike objects but evolved into fishlike objects for much the same fluid-dynamic reasons as those which caused fish to evolve into fishlike objects. Bicycles have evolved and so have motor cars. Even toothbrushes have evolved, though not very much. I have never seen Thomas Jefferson's toothbrush, but I don't suppose it was very different from the one we use today; the Duke of Wellington's, which I *have* seen, certainly was not. To some Victorian thinkers, facts like these served simply to confirm them in the belief that evolution was the fundamental and universal modality of change. The assimilation of technological to ordinary organic evolution was not wholly without substance, because all instruments that serve us are functionally parts of ourselves. Some instruments, like spectrophotometers, microscopes and radiotelescopes, are sensory accessories inasmuch as they enormously increase sensibility and the range and quality of the sensory input. Other instruments, like cutlery, hammers, guns and motor cars, are accessories to our effector organs – not sensory but motor accessories.

A property that all these instruments have in common is that they make no functional sense except as external organs of our own: all sensory instruments report back at some stage or by some route through our ordinary senses. All motor instruments receive their instructions from ourselves.

It was for reasons like this that the great actuary and demographer Alfred J. Lotka invented the word 'exosomatic' to refer to those instruments which, though not parts of the body, are nevertheless functionally integrated into ourselves. Everybody will have realised from personal experience how closely we are integrated psychologically with the instruments that serve us. When a car bumps into an obstacle, we wince as much from an actual referral of pain as from a sudden premonition of the sour and sceptical face of an insurance assessor. When the car is running badly and labours up hills, we ourselves feel rather poorly, but we feel good when the car runs well. Wilfrid Trotter, the surgeon, said that when a surgeon uses an instrument like a probe he actually refers the sense of touch to its tip. The probe has become an extension of his finger.

I do not think I need labour the point that this proxy evolution of human beings through exosomatic instruments has contributed more to our biological success than the conventional evolution of our own, or endosomatic, organs. But I do think it is worth while calling attention to some of the more striking differences between the two.

Genetic and exogenetic programmes

By far the most important difference is that the instructions for making endosomatic parts of ourselves, like kidneys and hearts and lungs, are genetically programmed. Instructions for making exosomatic organs are transmitted through non-genetic channels. In human beings, exogenetic heredity – the transfer of information through non-genetic channels – has become more important for our biological success than anything programmed in DNA. Through the direct action of the environment, we do in a sense 'learn' to develop a skin thicker on the soles of our feet than elsewhere. But information of this kind cannot be passed on genetically, and there is indeed no known mechanism by which it could be. It is only in exosomatic heredity that this kind of transfer can come about. We can learn to make and wear shoes and pass on this knowledge ready-made to the next generation as readily as we can pass on the shoes themselves.

There is no learning process in ordinary genetic heredity: we can't teach DNA anything, and there is no known process by which the translation of the instructions it embodies can be reversed. No information that the organism receives in its lifetime can be imprinted upon the DNA, but in exogenetic heredity we can and do learn things in the course of life which are transmitted to the succeeding generation; thus exogenetic heredity is Lamarckian or instructional in style, rather than Darwinian, or selective. By no manner of means can the black-smith transmit his brawny arms to his children, but there is nothing to stop his teaching his children his trade so that they grow up to be as strong and skilful as himself.

Learning as a new stratagem

The evolution of this learning process and the system of heredity that goes with it represents a fundamentally new biological stratagem – more important than any that preceded it – and totally unlike any other transaction of the organism with its environment. In ordinary, endo-somatic evolution and in cognate processes such as the so-called 'training' of bacteria and, in immunology, antibody formation, we are dealing with what are essentially *selective*, as opposed to instructive, phenomena. The variants that are proffered for selection arise either by some random process such as mutation or by a process which it is not paradoxical to describe as a 'programmed' randomness. By a 'programmed randomness' I mean a state of affairs in which the genera-tion of diversity is itself genetically provided for. Mendelian heredity provides for the preservation of genetic diversity for an unlimited period.

The reversibility of exogenetic evolution

Another important difference is this. Genetic evolution is conceivably reversible, just as it is thermodynamically conceivable that a kettle put on a lump of ice will boil. It's very unlikely, that's all. On the other hand, exosomatic evolution is quite easily reversible: everything that has been achieved by it can be lost or not reacquired. This is what specially frightens us when we contemplate the consequences of some particularly infamous tyranny that threatens to interrupt the cultural nexus between one generation and the next. This reversion to a cultural Stone Age is what each political party warns us will be the inevitable consequence of voting for the other. To bring the idea of reversibility to life, one should contemplate the plight of the human race if for any reason it did have to start again from scratch on a desert island: it is not heaven, but the Old Stone Age, that lies about us in our infancy.

Popper's third world

I have been looking around in my mind for some one word or phrase to epitomise what I understand by our human inheritance through non-genetic channels – through indoctrination, that is, and the conscious transfer of information by word of mouth and through books. Karl Popper's *Objective Knowledge*[1] supplied the answer ready-made. Let me therefore introduce you to Popper's concept of a 'third world'.

According to the philosophic views we specially associate with the name of George Berkeley, the apparently 'real' world about us exists only through and by virtue of our apprehension of it. Thus sensible things and material objects generally exist only as representations or conceptions or as 'ideas' in the mind – hence the name 'idealism'. Berkeley argued persuasively, but Boswell very well knew that Berkeley's argument was of just the kind that would enrage Dr Johnson. When Boswell teasingly said it was impossible to refute Berkeley's beliefs, 'I refute it *thus*', said Johnson, kicking a large stone so violently that he 'rebounded' from it, thus simultaneously refuting Berkeley and corroborating Newton's third law of motion (the one about actions' having equal and opposite reactions).

However, even those who take a sturdily Johnsonian view of Berkeley's philosophy as it relates to the real world of material objects sometimes hold a Berkeleian, or subjectivist, view of things of the mind. They tend to believe that thoughts exist by reason of being

[1] Karl R. Popper, *Objective Knowledge – an Evolutionary Approach* (Oxford, 1972).

thought about, conceptions by virtue of being conceived, theorems because they are the products of deductive reasoning, and beliefs because believed.

Popper's new ontology does away with subjectivism in the world of the mind. Human beings, he says, inhabit or interact with three quite distinct worlds: World 1 is the ordinary physical world, or world of physical states; World 2 is the mental world, or world of mental states; the 'third world' (you can see why he now prefers to call it World 3) is the world of actual or possible objects of thought – the world of concepts, ideas, theories, theorems, arguments and explanations – the world, let us say, of all artefacts of the mind. The elements of this world interact with each other much like the ordinary objects of the material world: two theories interact and lead to the formulation of a third; Wagner's music influenced Strauss's and his in turn all music written since. Again, I mention for what it may be worth that we speak of things of the mind in a revealingly objective way: we 'see' an argument, 'grasp' an idea, and 'handle' numbers, expertly or inexpertly as the case may be. The existence of World 3, inseparably bound up with human language, is the most distinctively human of all our possessions. This third world is not a fiction, Popper insists, but exists 'in reality'. It is a product of the human mind but yet is in large measure autonomous.

This was the conception I had been looking for: the third world is the greater and more important part of human inheritance. Its handing on from generation to generation is what above all else distinguishes man from beast.

Popper has argued strongly that, although the third world is a human artefact, yet it has an independent objective existence of its own – and is indeed quite largely autonomous. I have already pointed out that the third world undergoes the kind of slow, secular change that is described as evolutionary,[1] that is, is gradual, directional and integrative in the sense that it builds anew upon whatever level may have been achieved beforehand. The continuity of the third world depends upon a non-genetical means of communication and the evolutionary change is generally Lamarckian in character, but there are certain obvious parallels between exosomatic evolution and ordinary, organic

[1] *The Future of Man* (New York, 1959; London, 1960).

evolution in the Darwinian mode. Consider, for example, the evolution of aircraft and of motor cars. A new design is exposed to pretty heavy selection pressures through consumer preferences, 'market forces' and the exigencies of function, by which I mean that the aircraft must stay aloft and the cars must go where they are directed. A successful new design sweeps through the entire population of aircraft and cars and becomes a prevailing type, much as jet aircraft have replaced aircraft propelled by airscrews.

I hope it is not necessary to say that the secular changes undergone by the third world do not exemplify and are not the product of the workings of great, impersonal historical or sociological forces. Just as the third world, objectively speaking, is a human artefact, so also are all the laws and regulations which govern its transformations. The idea that human beings are powerless in the grip of vast historical forces is in the very deepest sense of the word nonsensical. Fatalism is the most abject form of the aberration of thought which Popper calls 'historicism'. Its acceptance or rejection has not depended upon cool philosophic thought but rather upon matters of mood and of prevailing literary fashion. There was quite a fashion for fatalism in late-Victorian and Edwardian England, admirably exemplified by Omar Fitzgerald's famous stanza:

> 'Tis all a Chequer-board of Nights and Days
> Where Destiny with Men for Pieces plays:
> Hither and thither moves, and mates, and slays,
> And one by one back in the Closet lays.

This is a comfortable doctrine, in so far as it spares us any exertion of thinking, but we may well wonder why it was so prevalent in England at that time.

This kind of fatalism sounds very dated now, but we should ask ourselves very seriously whether there is not a tendency today to take the almost equally discreditable view that the environment has already deteriorated beyond anything we can do to remedy it – that man has now to be punished for his abandonment of that nature which, according to the scenario of a popular Arcadian day-dream, should provide for all our reasonable requirements and find a remedy for all our misfortunes. It is this day-dream that lies at the root of today's rancorous criticism of science and the technologies[1] by people who believe, and seem almost to hope, that our environment is deteriorating to a level below which it cannot readily support human life. My own view is that these fears are greatly and unreasonably exaggerated. Our present dilemma has something in common with those logical paradoxes that have played such an important part in mathematical logic. Science and technology are held responsible for our present predicament but offer the only means of escaping their consequences.

The coming of technology and the new style of human evolution it made possible was an epoch in biological history as important as the evolution of man himself. We are now on the verge of a third episode, as important as either of these: that in which the whole human ambience – the human house – is of our own making and becomes as we intend it should be: a product of human thought – of deep and anxious thought, let us hope, and of forethought rather than afterthought. Such a union of the first and third worlds of Popper's ontology is entirely within our capabilities, provided it is henceforward made a focal point of creative thought.

The word 'ecology' has its root in the Greek word *oikos*, meaning 'house' or 'home'. Our future success depends upon the recognition that household management in this wider sense is the most backward branch of technology and therefore the one most urgently in need of development. An entirely new technology is required which is founded upon ecology in much the same way as medicine is founded on physiology. A blueprint for such a technology is described in the book *Only One Earth*, by Barbara Ward and René Dubos,[2] written in preparation for the United Nations World Conference on the Human Environment, held in Stockholm in 1972. If this new technology comes into being, I shall be completely confident of our ability to put and keep our house in order.

[1] P. B. Medawar and J. S. Medawar, in *Civilisation and Science: In Conflict or Collaboration?*, Ciba Foundation Symposium (London and New York, 1972).
[2] Barbara Ward and René Dubos, *Only One Earth: The Care and Maintenance of a Small Planet* (New York, 1972). See also pp. 279–86 below.

DOES ETHOLOGY THROW ANY LIGHT ON HUMAN BEHAVIOUR?

ALTHOUGH biologists take to it very kindly, the idea that the behaviour of animals can throw any light at all upon the behaviour of men is so far from self-evident that it would have been regarded with the utmost derision by the more tough-minded philosophers and philosophic thinkers of the seventeenth and eighteenth centuries – I mean, by men of the stature of Thomas Hobbes and Samuel Johnson.

Dr Johnson, in particular, would have been deeply outraged: 'Sir,' he would have said – he would certainly have said that, but what else he would have said can only be conjectural, though I think it might have run: 'is not the possession and exercise of moral judgement precisely the distinction between mankind and the brute creation? Show me an earthworm or marmoset that can tell the difference between right and wrong.' Thomas Hobbes would have pointed out – and in chapter 13 of *Leviathian* did, in effect, point out – that it is only by virtue of characteristics that differentiate civilised mankind from lesser beings that the life of man is anything but 'solitary, poor, nasty, brutish and short'.

Scientists of the twentieth century cannot be expected to quake in their shoes at the thought of the opinion that might have been held of them by eighteenth-century philosophers, however skilful and tough-minded they may have been, for they were simply not in possession of all the information that would have made it possible for them to form a definitive opinion. In particular, they were unaware of the evolutionary descent of man; but even if they had been apprised of it their reaction would probably have been 'What of it?' and Dr Johnson, wielding, as ever, the butt-end of his pistol, would have demanded to know what great and illuminating new truth about mankind followed from our realisation of his having evolved.

No one would answer this question today with anything like the blandly asseverative self-confidence that would have been characteristic of social Darwinists such as Francis Galton and Ernst Haeckel. To them it would have seemed obvious that mankind was stratified into superior and inferior human beings, and that all were engaged in a struggle for ascendancy – it would have seemed obvious, indeed, that 'man is a fighting animal'. It is, therefore, particularly satisfactory to put on record that it was, above all, ethologists who discredited the simple-minded and socially destructive beliefs characteristic of social Darwinism.[1] Some human beings are aggressive, to be sure, but it is by studying ourselves and not by studying animals that we recognise this trait in mankind; indeed, it is perhaps not unfair to say that those who know most about aggression in animals are most cautious in imputing any such thing as an aggressive 'instinct' to mankind.

This still leaves us, however, with the rather hectoring question of just what great new truth has been learnt about man as a result of the recognition of his evolutionary descent.

I now believe the question is wrongly put and that it embodies a false conception of the nature of scientific progress. All scientists despise the ideology of 'breakthroughs' – I mean the belief that science pro-ceeds from one revelation to another, each one opening up a new world of understanding and advancing still farther a sharp line of demarcation between what is true and what is false. Everyone actually engaged in scientific research knows that this way of looking at things is altogether misleading, and that the frontier between understanding and bewilderment is rather like the plasma membrane of a cell as it creeps over its substratum, a pushing forward here, a retraction there – an exploratory probing that will eventually move forward the whole body of the cell.

It is, indeed, not a grand ethological revelation that the scientist should seek from his awareness of the evolutionary process, but rather an enlargement of the understanding made possible by a new or wider angle of vision, a clue here and an apt analogy there, and a general sense of evolutionary depth in contexts in which it might otherwise be lacking. In such an exercise as this, as Robert Hinde[2] points out, we

[1] e.g. Crook and others in M. F. A. Montague (ed.), *Man and Aggression* (New York and London, 1968); R. A. Hinde, *Animal Behaviour: A Synthesis of Ethology and Comparative Psychology* (New York, 1970).

[2] R. A. Hinde, 'The Use of Differences and Similarities in Comparative Psychopathology', in George Serban and Arthur Kling (eds), *Animal Models in*

often have as much to learn from the differences as from the similarities between human beings and animal models. In short, I think ethology is one of the many areas of thought in which a philosophic understanding of the nature of the scientific process is salutary: in real life, science does not prance from one mountain top to the next.

A great deal obviously depends upon whether or not it is possible to establish genuine homologies between the behaviour of human beings and that of the collateral descendants of their remote ancestors. To my mind, there can be no question that we can do so: consider, for example, the highly complex and teleonomically related behavioural and physiological activities that go to make up sexually appetitive behaviour, mating, conception, gestation, parturition, lactation, suckling and the care of the young. It is hardly conceivable that this entire complex scenario should have sprung into being for the first time with the evolution of *Homo sapiens*; and there is no reason nowadays why we should put ourselves to the exertion of imagining that it could have done so. The same applies to appetitive behaviour as it relates to the conventional everyday context of seeking food.

Nevertheless, even if the existence of these homologies is conceded, we must not expect too much to follow from them; in particular, we should not expect a great variety of psychologically illuminating insights, for ethology often stops short at just the level at which psychology begins, that is, stops short of explaining the nature and origin of the *differences* between individuals. As I pointed out in an early essay on this very subject,[1] what is *interesting* about our human propensity for loving or hating is why one person loves a second and hates a third, just as what is interesting about our human dietetic habits is not the physiological need for food or the behaviour that goes with it, but why any one human being will eat this and not that, here and not there, and now and not then. Psychology has had to do with the differences between human beings, ethology – at least in its early days – with the characteristics they have in common. I went on to say: 'It is no great new truth that human beings are ambitious; what

---

Human Psychobiology, Proceedings of the Second International Symposium of the Kittay Scientific Foundation, 1974 (New York and London, 1976).

[1] 'The Uniqueness of the Individual', in *The Uniqueness of the Individual* (London, 1957; New York, 1981).

is interesting about ambition is why in one person it should take the form of wanting to become a great musician and in another of wanting to raise a large family, and in a third (for this, too, is an ambition) of wanting to do nothing at all.' These questions seem to me to belong to the domain of psychology and not to ethology at all.

I do not think there is any general answer to the question of how exact are the homologies between, say, human and primate behaviour. Scientifically, it would be completely sterile to start with the presumption that homologies were illusory and the correspondences invariably inexact. Methodologically, the sensible thing is to start with the hypothesis that the homologies are fairly close; for such a hypothesis can be made the basis of action and can put us on the right wavelength for making observations of the kind which will either falsify it or strengthen our belief that there may be something in it. For example, if it were to be shown that maternal deprivation had a psychologically damaging effect on rhesus monkey babies, then there would be a case prima facie for supposing that the same might be true of human beings. If the idea is wrong it will be shown to be so, for it is the great and distinctive strength of science that we need not persist long in error if there is a genuine determination to expose our ideas to tough critical analysis. In general, it is illuminating to recognise that such human – sometimes all too human – activities as play, showing off and sexual rivalry are not psychic innovations of mankind, but have deep evolutionary roots.

I mention it purely as an aside that the ancient origins and deeply programmatic character of human reproductive behaviour makes it extremely unlikely that a prudential system of sexual morality can be reinstated or the population explosion contained merely by exhortation or appeals to reason, however cogently worded. It also makes it extremely unlikely that there is any easy psychological solution of the problem of family limitation. On the subject of behavioural homologies as they apply to man, I cannot do better than to quote the fine closing paragraph of Darwin's *The Descent of Man*:[1]

We are not here concerned with hopes or fears, only with the truth as far as our reason permits us to discover it, and I have given evidence to the best of my

[1] C. Darwin, *The Descent of Man* (London, 1871).

ability. We must, however, acknowledge as it seems to me, that man with all his noble qualities, with sympathy which he feels for the most debased, with benevolence that extends not only to other men but to the humblest living creature, with his godlike intellect which has penetrated into the movements and constitution of the solar system – with all these exalted powers – man still bears in his bodily frame the indelible stamp of his lowly origin.

For our own purposes the quotation must, of course, be modified so as to read 'behavioural repertoire' instead of 'bodily frame'. With this modification, the only word we could reasonably cavil at is the word 'indelible', for we are by no means at the mercy of our biological inheritance. For example: human evolution has taken such a course that haemolytic disease of the newborn is almost inevitable in certain situations unless we take steps to circumvent it, but we *can* and do take steps to do so. The same applies to human behaviour: our social conventions and institutions do inhibit and to some extent prevent the more exuberant forms of the behaviour so often described as 'bestial' though in reality it is more characteristic of man than of beast.

I should like to quote what has always seemed to me to be a very wise passage by Alfred North Whitehead in his *Introduction to Mathematics*:[1] 'It is a profoundly erroneous truism, repeated by all copybooks and by eminent people when they are making speeches, that we should cultivate the habit of thinking of what we are doing. The precise opposite is the case. Civilisation advances by extending the number of operations which we can perform without thinking about them.' I find this statement compellingly true of much human behaviour that can be described as skilled, for if somebody asks us what seven sixes are (or anything else from our times-tables) the last thing on earth we want to do is to go back to Peano and Frege or to Russell and Whitehead to puzzle out what the answer to such a searching question can be: we want to snap it out without hesitation; and likewise the process of learning to drive a car, obeying traffic signals and avoiding obstacles etc., is in effect learning *not* to think about actions which did, at one time, require anxious deliberation ('Now let me see, which of these many pedals will arrest the motion of the car?'). We are learning to give learned behaviour the polish and fitness for purpose which we describe

[1] A. N. Whitehead, *An Introduction to Mathematics* (London, 1911).

colloquially as instinctive, so that we can drive a car safely even if we are tired, preoccupied and in a hurry. In the same essay I wrote:

Paradoxically enough, learning is learning *not* to think about operations that once needed to be thought about; we do, in a sense, strive to make learning 'instinctive', i.e. to give learned behaviour the readiness and aptness and accomplishment that are characteristic of instinctive behaviour. [*But that is only half the story*.] The other half of the half truth is that civilization also advances by a process that is the very converse of that which Whitehead described: by learning to think about, adjust, subdue and redirect activities which are thoughtless to begin with because they are instinctive. Civilization also advances by bringing instinctive activities within the domain of rational thought, by making them reasonable, proper and co-operative. Learning, therefore, is a twofold process: we learn to make the processes of deliberate thought 'instinctive' and automatic, and we learn to make automatic and instinctive processes the subject of discriminating thought.

Anybody who professes to discern a moralising flavour in what I have been writing is perfectly right: it is exactly what I intend. I think we shall have to get used to the idea that moral judgements should intrude into the execution and application of science at every level – and in no context more exigently than the interpretation of science for the benefit of laymen.

Anyhow, these reflections on Whitehead lead me naturally to one or two points I should like to make on the special importance of language and the learning process in human beings and their relevance to the conception and measurement of 'intelligence'.

Having dismissed the idea that we must look to ethology for great revelations about the nature of human behaviour, and also the idea that scientific understanding lurches forward from one revelation to another, I should like to put on record my belief that the great contribution of ethology to the understanding of human behaviour is methodological: indeed, there are already signs that some of the more eclectic psychiatrists, anxious as ever to be in the swim and in the forefront of opinion, are starting to speak about ethology as if they had invented it themselves. This will be all to the good if it has the effect of persuading them to direct upon human beings the intent and candid gaze to which ethologists owe their success in the study of animal behaviour.

The very last thing I have in mind is anything like the at one time popular sociological exercise known as 'mass observation' – a kind of inductivism gone mad, as I remember it – which was dedicated to the proposition that if only one could collect enough information about what people actually do and actually say at home, in pubs and in buses etc., then some great new truth about human behaviour would, of necessity, emerge. So far as I am aware, no truth of any kind emerged, great or small, except perhaps that it takes all sorts to make a world. No: I was allowing myself to cherish the hope that ethological methods might one day make it possible to build up a biologically well-founded psychology or even a psychopathology to take the place of the weird farrago of beliefs which forms the basis of modern psychoanalytic psychotherapy, a system of beliefs which persists because it has never been found wanting, and has never been found wanting because it has never been exposed to any evaluation.

I have defended on several occasions[1] the view that what is characteristic of human beings considered as animals is not the making and use of tools but the communication from one person to another, specially in the next generation, of the knowledge and know-how required to do so. It is by virtue of this faculty that human beings come to enjoy a kind of cultural evolution which has converted us into animals that are simultaneously aerial, terrestrial and submarine, possessing X-ray eyes and sense organs sensitive enough to feel the heat of a candle at a distance of a mile. In briefest summary, because there is no point in going over old ground again, the main characteristics of this distinctively human form of evolution are:

1   It is Lamarckian in style, that is, unlike ordinary or 'endosomatic' evolution it embodies a learning process; for what is learned is passed on and becomes part of the evolutionary heritage.

2   Cultural heredity is mediated through non-genetic channels – hence the use of the terms 'exogenetic' or 'exosomatic' evolution to describe the process and distinguish it from ordinary genetic or Darwinian evolution.

3   It is reversible. Not just in theory – as it is possible in theory

---

[1] *The Uniqueness of the Individual* (see p. 193 above, note 1); *The Future of Man* (see p. 188 above, note 1); 'Technology and Evolution', pp. 184–90 above.

anyway for a kettle to boil when it is put on ice – but in good and earnest. We could return to the Stone Age in one generation.

This system of evolution is the characteristic to which we owe our clear-cut biological supremacy over all other organisms, because it has conferred almost unlimited capabilities upon us, including even that of leaving the earth and going to live elsewhere in the solar system.

This characteristically human system of heredity calls for and depends upon the existence of language and other forms of conceptual communication (one is reminded that Dr George Steiner refers to man as a 'language animal' rather than as a 'tool-making animal'). The existence of and our dependence on exogenetic evolution must place a specially high selective premium upon such capabilities as teachability and imitativeness (a word for which, in this context, the vernacular term 'aping' seems uncannily apt) because these form the causal nexus of cultural heredity – points made very well by Barnett in a recent lecture entitled '*Homo docens*'.[1]

With this system of heredity in mind the distinction between programmed and learned behaviour loses some of its force, for an episode of behaviour can be 'programmed' in the sense that the physical plant and the functional capabilities necessary for its execution must be ready-made even though the behaviour itself has to be learned. Barnett himself attaches considerable importance to the observations of Kuo[2] on the killing of rats and mice by cats. It is clear that although a cat's aptitude and general capability for killing mice is 'laid on' developmentally, cats are much more likely to kill mice if they have observed their mothers doing so.

The authors surmise that this kind of behaviour may be characteristic of Felidae generally, noting, however, that no experiments on the subject have yet been done on tigers or lions; but if we reflect on the importance attached to teaching and the entire apparatus of pedagogy in human beings it seems difficult to resist the hypothesis that the same principle is true of human beings.[3]

[1] S. A. Barnett, '*Homo docens*', *Journal of Biosocial Science*, **5**, 393–403, 1973.

[2] Z. Y. Kuo, 'Genesis of the Cat's Behaviour Towards the Rat', *Journal of Comparative Psychology*, **11**, 1–35, 1930.

[3] M. E. P. Seligman and J. L. Hager (eds), *Biological Boundaries of Learning*

In the light of these considerations I wonder increasingly at the *naïveté* of those psychologists who believe that prowess in certain so-called 'intelligence tests' provides a measure of a so-called 'innate intelligence' which is virtually unaffected by the subject's age and which cannot be taught or influenced by experience. If this were true of any human performance, my first reaction would be to dismiss that performance as relatively unimportant and certainly not one that could be made the basis of a measure of intelligence, for I believe that the endowments that have made human beings what they are, are above all imitativeness and teachability.[1]

For these and other reasons I believe that the IQ concept and some of its practitioners should now be relegated to that dusty, cavernous and ill-lit building called the Museum of Social Darwinism, in which other principal exhibits are Francis Galton, Ernst Haeckel and Alfred Rosenberg, and many others who have misunderstood the bearing of biology on human affairs or who have propagated mischievous views in the name of science.

POSTSCRIPT

When the conference at which I read this paper was planned, the title I had submitted for a contribution that I thought would come well from a general biologist was: 'Does the study of the behaviour of animals throw any light on the behaviour of Man?', and this is the topic on which I wrote my paper. The title was such a mouthful, however, that I changed it to its present form.

Put colloquially, my answer to the original question can be seen from the text to be: 'Not very much, really – it is by studying human beings themselves that we learn about their behaviour.' If, however, we put the question in the form in which it actually appears at the head of my contribution, then I think the proceedings of the conference make it possible to say with some confidence: 'Yes, indeed.' This puts me

---

(New York, 1972); R. A. Hinde and J. Stevenson Hinde (eds), *Constraints on Learning: Limitations and Predispositions* (London and New York, 1973).

[1] These views were the subject of an altercation between Professor H. J. Eysenck and myself in a number of issues of the *New Statesman and Nation* (11 January 1974 and two succeeding issues).

under an obligation to try to describe what differentiates ethology from
psychology in a conventional sense.

I think ethology has two distinctions: in trying to make teleonomic
sense of behavioural performances that might seem to inexperienced
observers to be a stream of incoherent and functionless activities,
ethologists are not yet importuned by an insistent and urgent need to
find a causal explanation for every phenomenon they observe. Closely
related to this is the welcome truth that ethology, unlike some psychological systems, is not yet crabbed and confined by the doctrinal
tyranny of any pre-existing explanatory system. These two characteristics give ethology the freshness and spontaneity which other biologists
find so enviable, and which are sadly so lacking from many of the
older and more conventional branches of zoology.

I turn now to the consideration of a number of comments and criticisms that have been made about the first version of this paper.

Apropos of exogenetic evolution, Professor Niko Tinbergen, who
was unfortunately unable to be present at the meeting, expressed some
doubt about the total reversibility I claimed for it, and in particular
about my contention that human beings could revert to the Stone Age
in one generation if the cultural nexus between one generation and the
next were to be wholly severed. The matter can only be resolved by a
purely notional experiment in which we are to imagine the development and likely fate of a community consisting of a number of human
babes reared as if by magic on a desert island – kept alive, indeed, but
without the benefit of any of that inheritance which is passed on from
one generation to the next by precept, books or word of mouth.
Without that inheritance – without any formal schooling or indoctrination in the useful arts – would they not regenerate among themselves
into something like a palaeolithic culture?

This experiment becomes more and more vague and unsatisfactory
the more one thinks about it, so there is no point in going on about it.
I admit, though, that the 'Stone Age' was a rhetorical over-simplifica-
tion; the point I really want to make, which I think most people concede, is that exogenetic evolution is *reversible*.

Another point made by Tinbergen was that we should seek analogies
no less often that homologies – meaning by analogies 'similarities con-
vergently evolved in less closely related species in adaptation to similar

niches'. I accept this criticism completely, and smiled at the justice of the rebuke given to someone trained in a school of zoology for which the notion of 'homology' was the central – even the energising – concept of the whole of biology.

I am not very well up in the literature of ethology and was not therefore surprised to learn several examples unfamiliar to me of exogenetic evolution or cultural transmission among animals. Tinbergen, in particular, referred me to 'acculturation' in Japanese macaques and referred me to a passage by Hinde;[1] in addition, Mr Gady Katzir called my attention to a description of a cognate phenomenon by Marais.[2]

Amid the general discussion that followed my paper I reminded the members of the symposium that molecular genetics began with the complete solution, by intent and single-minded research, of the phenomenon of pneumococcal transformation, and I went on to ask if there were not some comparable phenomenon in ethology, the complete interpretation of which would have an effect analogous to that of an aircraft's breaking cloud; I referred to the transformation of male into female sexual behaviour and vice versa by the injection of hormones characteristic of the opposite sex – phenomena of which a number of specially apt examples were given by Lehrman.[3] A complete interpretation of any one of these phenomena, I argued, might have an effect on ethology comparable with the effect of O. T. Avery's work on the subsequent growth of molecular genetics. In the discussion which followed it became clear that members of the conference were not at all sure that any such parallel between ethology and the growth of molecular biology could be drawn.

It may, in any case, turn out that the revolution I have in mind will be conceptual rather than one that turns upon the analysis of behavioural phenomena. Commenting on my paper both Hinde and Bateson professed the rather strong feeling that the Lorenzian distinction

[1] R. A. Hinde, *Biological Bases of Human Social Behaviour* (New York, 1974), pp. 243–4.
[2] E. Marais, *The Soul of the Ape* (London, 1969; 2nd ed., Harmondsworth, 1973).
[3] D. S. Lehrman, 'Gonadal Hormones and Parental Behaviour in Birds and Infra-human Mammals', in W. C. Young (ed.), *Sex and Internal Secretion* (Baltimore, 1961).

between instinctual and learned behaviour is no longer useful: 'all behaviour is both genetically and environmentally influenced; all learned behaviour is genetically influenced'.

Methodologically speaking, I think this is a rather weak declaration and part of it, at least, is tautologous. Obviously no behavioural performance could take place without the physical and physiological means to execute it – there must be nerves and muscles and other apparatus of the kind we normally think of as being 'laid on' by developmental processes. It could be said of *every* character trait whatsoever that its determination is partly natural and partly nurtural, yet we do know of character differences that are wholly genetic in determination in the usual sense of the term – for example, a human being carrying the blood-group gene associated with group A will be of group A in almost any environment that is capable of sustaining life. It is an old, vexed question, and it will not be solved in the context of behaviour until Aubrey Manning's ambition for the foundation of the genetics of behaviour is realised.

TAKING THE MEASURE OF MAN

I⟮ is ironic that Karl Pearson's great monument to Galton – his copiously detailed and reverential *The Life, Letters and Labours of Francis Galton* – has probably deterred many would-be biographers. D. W. Forrest has fortunately not been deterred, and although he acknowledges his indebtedness to Pearson he is well aware of the weakness of the kind of biography of which Dr Johnson said that it is difficult to discern the outline of a man through the mist of panegyric. Professor Forrest's admirable biography[1] is a good deal more ambitious and wider-ranging than C. P. Blacker's short but very readable *Eugenics: Galton and After*.[2] Professor Forrest is not reverential, though he is not always adequately critical, and the picture he paints is not more than life-size – a difficult feat when writing of a man of Galton's audacity, versatility and enormous intellectual appetite: it must have been very difficult to avoid going on and on about some of the more interesting of Galton's activities. This danger has been avoided and the text is a rounded whole of reasonable size and proportions.

Professor Forrest is about as authoritative as any one man could be in the appraisal of activities as diverse as Galton's. It would need a committee to weigh up his contributions to meteorology (the term 'anticyclone' is Galton's), to statistics (particularly in the development of correlation theory), to identification by fingerprints – and, equally important, the identification of the fingerprints themselves by a judicious taxonomy; to psychology, exploration and anthropometry; especially to eugenics (Galton was its chief spokesman and coined the name that describes it), and to genetical methodology as well – particularly the study of twins to distinguish between the contributions of nature and nurture to the differences between human beings. Galton devised stereophotographic maps and made much use of composite photographs for anthropological purposes.

[1] D. W. Forrest, *Francis Galton: The Life and Work of a Victorian Genius* (London, 1974).
[2] (London, 1952).

One of the attractions of Professor Forrest's biography is a number of amusing and informative digressions prompted by the fact that Galton's maternal grandfather was Erasmus Darwin. We learn, for example, that it was Pitt's Under-Secretary for Foreign Affairs, Canning – no less – whose lampoon *The Loves of the Triangles* put an end to Erasmus Darwin's poetic pretensions. We learn also that Erasmus Darwin the physician sometimes charged fees of forty or even a hundred guineas, the latter for attendance on a feverish child (a pretty steep fee, too, having regard to the fact that he had no specific remedies for the infections of which these fevers were almost certainly the consequence).

Nothing is more characteristic of Galton's genius than his great ingenuity in devising statistical measures of human faculties and traits of character: he invented means of measuring good-temperedness versus bad-temperedness, boredom versus attention, the efficacy of prayer, and much else besides. He loved measuring and conducting statistical exercises founded on these measurements, but I fear that this predilection may have been his undoing as a geneticist. Galton thought of his genetical writings as so many prolegomena to a grand Theory of Heredity, which fortunately he never reached the stage of formulating, for he was altogether too ready to believe that such a theory could be propounded by purely ratiocinative methods – by the examination and analysis of measurements and their correlations – without recourse to experimentation in the Baconian sense.

Galton's work reached its full development in Karl Pearson's research on heredity. This makes it possible to say with confidence that Galton as a geneticist was working and thinking on the wrong lines, for during his own lifetime the history of science had quietly taken a quite different turning: the work of Mendel and later of Bateson and Punnett led to the institution of the science Bateson called 'genetics', a body of theory about the origin and distribution of differences between individuals. Galton is, of course, no more to be reproached than is Darwin for having worked in ignorance of discoveries that had not yet been made. Although he was ahead of his contemporaries in recognising the combinatorial nature of the differences between individuals, he never effectively realised that for this reason he should study the inheritance of quantal differences; but in his

Natural Inheritance he shows excellent judgement in repudiating the doctrine that has since come to be called Lamarckism (the belief that characteristics acquired in an organism's lifetime may become part of its hereditary make-up).

Galton was a serious traveller and anthropological investigator, and Professor Forrest cites passages from his accounts of his travels which, as well as being amusing, illuminate the experiences that must have helped to shape Galton's later views on the relative merits of different races.

In south-west Africa his transactions with the tribal chief Nangoro led him from one social blunder to another: 'When invited to a meal Galton refused to take part in a cleansing ritual in which the host gargled water and ejected it over the guest's face. As this technique had been devised by Nangoro himself as a countercharm against witchcraft Galton's unclean presence at the table led to some constraint.' Worse still, he bundled out of his tent a princess presented to him as a concubine, despite the enhancement of her charms by a liberal smearing of red ochre and butter.

Galton's great contributions to psychology were psychometry, the statistical techniques needed to make use of the information it yielded, his advocacy of the use of twins in research, his just and firmly declared conviction that differences of intelligence were inherited – a belief in which he carried with him Darwin (who mentions him more than once in *The Descent of Man*) – and his total repudiation of the idea of the equality of man and with it the belief that a man is only what his environment and upbringing can make of him.

There is a juxtaposition of passages early in Professor Forrest's book which struck me as very significant for understanding Galton as a psychologist, though I am quite sure that the author did not intend their conjunction to have the significance I read into them. On page 7 he cites Terman's estimate of Galton's IQ – 200 – a figure of which Terman says that it was not equalled by more than one child in 50,000 of the generality; on page 8 a passage quoted from Galton's memoirs calls attention to his wonderment, at the age of eight, that the copies of Caesar's commentaries distributed for class use should still be so fresh and clean considering that they were contemporaneous with Caesar himself and thus nearly 2,000 years old. It is possible, then, that Galton was a forerunner of modern IQ psychologists in combining (as I believe they often combine) high intelligence, in the sense represented by a proficiency in performing computer-like intellections, with a relative lack of intelligence as a humanist understands the term: a combination of strong understanding and commonsensical judgement.

Galton's psychology and his eugenics were closely intertwined: his opinions are those of an ostensibly sensible, well-off, right-thinking upper-middle-class professional man with a strongly conservative temperament. He accepted as a matter of course the opinions of the England of his day.

It is almost but not quite unnecessary to say that Galton disapproved of the higher education of women. He warned Professor Thorndike of Columbia University that higher education would either select those who had a small probability of marriage or would bring out in them qualities which would lessen that probability. Galton joined the Anti-Suffrage Society and held the, at that time, conventional belief that women were intended and fit only for reproduction.

On the credit side, however, it should be recorded that Galton resigned from the Royal Geographical Society when its Council disregarded the wishes of the Fellows (secured by referendum) and managed to negative the motion that women should be elected as Fellows. Doubtless they were very much influenced by the authoritative opinion of Curzon, the future Viceroy of India, who contested 'in toto the general capability of women to contribute to scientific geographical knowledge', adding that American women globe-trotters were one of the horrors of the latter end of the nineteenth century. A nephew says of Galton that the only man of whom he ever heard him speak harshly was Oscar Wilde.

Professor Forrest makes the case for Galton's being rated a genius admirably well, and he shows him as a kindly and by no means disagreeable man; it is fair to concede that many attacks upon Galton as a eugenist are sometimes founded not so much upon what he said as upon a certain conventional misrepresentation of his views. Nevertheless, there is a side of Galton's character that looks very disagreeable to modern eyes. For example, in *Hereditary Genius*[1] there is an air of almost exultant scorn in his description of the uselessness of a man's

[1] (London, 1869), pp. 14–16.

trying to better himself beyond the degree fixed by his innate capabilities. Such a man, he says, will find himself at whatever station of life it may have pleased nature to allot him, whereupon, if he has any judgement, he is no longer tormented into hopeless efforts by the fallacious promptings of vanity, and finds true moral repose in the honest conviction that he is engaged in work as good as nature has rendered him capable of performing. No one acquainted with the biographies of the heroes of history could doubt the existence of grand human animals, of individuals born to be kings of men, but middling achievement was to be the lot of most people. It is Nietzsche's slave morality served up as science.

Again, a passage in *Natural Inheritance* shows how closely Galton's scientific opinions were intertwined with his political outlook: he points out how meagre is the information given by the average value of some variable quality and insists upon the importance of measuring racial or popular *distribution*, for 'a knowledge of the distribution of inequality enables us to ascertain the Rank that each man holds among his fellows in respect of that quality. This is a valuable piece of knowledge in this struggling and competitive world in which success is to the foremost and failure to the hindmost irrespective of absolute efficiency.'

In fact, the idea that some of Galton's thinking reveals him as a spiritual Fascist cannot be dismissed as an unfounded calumny. Galton envisaged the privilege of reproduction's being enjoyed by a 'gifted class' who would have a preponderant say in legislation, particularly in laws affecting the inheritance of wealth; Galton said he did not see why insolence of caste should prevent their treating their inferiors kindly so long as they remained childless; but, he went on to say, if they were to continue to procreate children inferior in moral, intellectual and physical qualities it was easy to believe that the time might come when such persons would be considered enemies of the State and to have forfeited all claims to kindness.

Needless to say, Professor Forrest does not applaud these views and it is characteristic of the general tact and discretion of his biography that he clearly regards exposition as condemnation enough; and so it is, for there is an intolerable stench of the gas chamber about the whole idea.

In spite of all his imperfections, some of them grave, and the confinement of his thought by the prejudices of his day, Galton's great and abiding achievement was to have brought into the domain of scientific enquiry a great many subjects thought until then to lie outside it. His work had a seminal influence on psychology and he is a most important figure in the history of ideas generally.

HERBERT SPENCER AND THE
LAW OF GENERAL EVOLUTION

I HESITATED some little while before accepting Oxford's invitation to deliver a Herbert Spencer Lecture in 1963. My conscience told me that I could not honourably accept unless I were prepared to steep myself in the work and thought of Spencer himself, an enterprise not to be lightly undertaken. For Spencer was a system philosopher, one who endeavoured (in Whitehead's words) 'to frame a coherent, logical, necessary system of general ideas in terms of which every element of our experience can be interpreted'.[1] And his system was set forth in twelve volumes thicker and squarer than Gibbon's, each bound in a cloth which has acquired with age a reptilian colour and texture, so putting one in mind of some great extinct monster of philosophic learning.

The prospect did not beckon me on; nor was I cheered up by finding myself the first man ever to have read the two volumes of the *Principles of Biology* acquired by the Royal Society's library more than half a century beforehand. But when I began to read I knew that the challenge was not to be resisted. I began to understand why in his lifetime Herbert Spencer's work had sold in tens of thousands. The *Study of Sociology* at 10s. 6d. had sold more than 20,000 copies by 1900, and the cheap edition of his tract on *Education* nearly 50,000 – and these, it should be remembered, lay outside his formal *System of Synthetic Philosophy*. The system comprised the principles of sociology, and of psychology, biology and ethics, the whole knit together by a volume of *First Principles*, 'primordial truths' arrived at by deduction from 'the elementary datum of consciousness'. What a tremendous undertaking! And what a formidable man Spencer was! His energy was apparently equal to any exertion; his thought had a steady pounding forward motion along the lines he had laid down for it in the famous manifesto or prospectus that preceded the first edition of the *First Principles*.[2]

[1] A. N. Whitehead, *Process and Reality* (Cambridge, 1929).
[2] *Le style est l'homme même*. I had quite forgotten, when I wrote this, that Spencer began his professional life as a railway engineer.

I think Spencer was the greatest of those who have attempted to found a metaphysical system on naturalistic principles. It is out of date, of course, this style of thought; it is philosophy for an age of steam; and until a few years ago we should have been tempted to describe it as equally out of date in content. Evidently it is not. Spencer's greatest contribution to philosophy was his theory of general evolution, which I shall expound in a moment, and in recent years his ideas have come back to life, or been propped upright again, in the work of men as far apart as Julian Huxley and Teilhard de Chardin,[1] to say nothing of the revival of evolutionary sociology and social anthropology. I intend it to be a compliment to both parties when I say that Huxley's thought about evolution is in the same general style as Spencer's. Teilhard, on the contrary, was in no serious sense a thinker. He had about him that innocence which makes it easy to understand why the forger of the Piltdown skull should have chosen Teilhard to be the discoverer of its canine tooth.[2]

Now what I propose to do is to show that the principle of general evolution is not an important principle, and that abstractions arrived at in the way this one was arrived at have the property of sounding as if they were tremendously 'significant' without actually being so. I then turn to a much more important problem arising directly out of Spencer's evolutionism, namely the difficulty he felt obscurely (and others have since felt very clearly) of reconciling the law of general evolution with another great natural law – one that pronounces for a general decay of order and a great levelling of energy, and declares that the direction of the flow of natural events is always towards what Willard Gibbs called *mixedupness*. I shall try to identify the various misunderstandings which have led to the belief that living organisms circumvent or actually break this second law of thermodynamics, and shall then argue that the equation of biological order or organisation with thermodynamic order, and so in turn with information content and the idea of improbability, is one that cannot be sustained.

To show that I have no grudge against evolution as such, and that I

---

[1] See *The Phenomenon of Man* (London, 1959), reviewed below, pp. 242–51.

[2] On 30 August 1913. The whole story is to be found in *The Earliest Englishman* by A. S. Woodward (London, 1948); see also Charles Dawson and A. S. Woodward, *Quarterly Journal of the Geological Society*, **70**, 82–99, 1914.

accept the laws of thermodynamics in the spirit in which Carlyle's lady correspondent accepted the universe ('by Gad she'd better!'), I emphasise that I shall say nothing about the principle of general evolution that I would not be prepared to say about any other attempt to pass off a mere inductive *collage* as a work of philosophic art.

Consider, for example, a great new universal principle of complementarity (not Bohr's) according to which there is an essential inner similarity in the relationships that hold between antigen and antibody, male and female, electropositive and electronegative, thesis and antithesis, and so on. These pairs have indeed a certain 'matching oppositeness' in common, but that is *all* they have in common. The similarity between them is not the taxonomic key to some other, deeper affinity, and our recognising its existence marks the end, not the inauguration, of a train of thought. The several manifestations of complementarity are so completely different in origin, nature and import that the properties of the one pair need teach us nothing about the properties of any other. What we do learn is to recognise the relationship when it turns up in a new and unfamiliar context. The idea of complementarity has, for example, never been far from the thoughts of those who have tried to find out how two chromosomes come to be formed where there was only one before; and for biologists the quintessential example of complementarity is indeed the relationship between the twin strands of the molecule of deoxyribonucleic acid, DNA.

Spencer was the first great evolutionist, and he gave the word *evolution* its modern connotation in English. His first account of the matter is in his 'Development Hypothesis',[1] a (for Spencer) relaxed and fairly chatty argument that appeared in the *Leader* between 1851 and 1854; that is, seven years before the publication of the *Origin of Species*, and when Spencer himself was in his early thirties. In it Spencer asks why people find it so very difficult to suppose 'that by any series of changes a protozoon should ever become a mammal' while an equally wonderful process of evolution, the development of an adult organism from a mere egg, stares them in the face. We can tell from the tone of his article that evolution was already an idea widely discussed by people of philosophic tastes.

[1] Reprinted in *Essays: Scientific, Political and Speculative* (London, 1868).

As his thought developed, Spencer came to think of genetic evolution, evolution in Darwin's sense, as no more than one manifestation of a far grander and more pervasive process; and out of this conviction his system grew. Today we realise that philosophers devise systems because it gives them a nice warm comfortable feeling inside; it is something done primarily for their benefit, not for ours. Spencer would not have taken kindly to such an interpretation. Nor did he believe that his concept of general evolution grew inductively out of the contemplation of its several instances. On the contrary: Spencer, like Whitehead after him (the last of the great system-philosophers), undertook 'the deduction of scientific concepts from the simplest elements of our perceptual knowledge'.[1] *First Principles* was an attempt to do just this: to show that the concept of general evolution followed 'inevitably' from laws of the indestructibility of matter and of the conservation of energy. Spencer's argument is unimportant and unconvincing, its sole purpose being to justify his expectation of finding evolution at work everywhere. The universe evolved, and the solar system and earth within it. Animals and plants evolve generation by generation, and within any one generation the development of each individual is itself an evolution. Society is an organism and society evolves. Moreover, 'the law of evolution holds of the inner world as it does of the outer world'. Mind evolves; and language and musical expression, the plastic arts and the arts of narrative and dancing, all display one characteristic or another of evolutionary change. Evolution is 'a universal process of things'. And when we contemplate it as a whole, in its 'astronomic, geologic, biologic, psychologic, sociologic, etc.' manifestations, 'we see at once that there are not several kinds of Evolution having certain traits in common, but one Evolution going on everywhere after the same manner . . . So understood, Evolution becomes not one in principle only but one in fact.'[2] These larger ideas, I should explain, grew upon Spencer during the latter part of his life. I am quoting from the last edition of *First Principles*; they are not be to found in the first edition of 1862.

What then was this universal law of the transformation of matter

[1] A. N. Whitehead, *An Enquiry concerning the Principles of Natural Knowledge* (Cambridge, 1919).
[2] *First Principles*, 6th ed. (London, 1900), § 188.

and energy? He picked his way towards a definition or description that satisfied him, but even after forty years he was wanting to polish and qualify it still. What he has to say about definition itself, the process of defining, is an example of his splendid good sense and of his powerful, hideous prose – the writing of a man who, lacking and perhaps contemptuous of the stylistic graces, is absolutely determined to be understood:

A preliminary conception, indefinite but comprehensive, is needful as an introduction to a definite conception. A complex idea is not communicable directly, by giving one after another its component parts in their finished forms; since if no outline pre-exists in the mind of the recipient these component parts will not be rightly combined. Much labour has to be gone through which would have been saved had a general notion, however cloudy, been conveyed before the distinct and detailed delineation was commenced.

The point is commonplace nowadays, but many scientists still persist in the belief that no rational discourse is possible unless one 'defines one's terms'.

In the outcome, Spencer's definition, as it is to be found in the final revise of *First Principles*, ran thus:

Evolution is an integration of matter and concomitant dissipation of motion; during which the matter passes from an indefinite, incoherent homogeneity to a definite, coherent heterogeneity; and during which the retained motion undergoes a parallel transformation.

At once he goes on to say, and we love him for it:

NOTE. – Only at the last moment, when this sheet is ready for press and all the rest of the volume is standing in type . . . have I perceived that the above formula should be slightly modified . . . by introduction of the word 'relatively' before each of its antithetical clauses.

What his principle of general evolution amounts to is this: that the direction of the flow of events in the universe is from simple to complex, diffuse to integrated, incoherent to coherent, independent to interdependent, undifferentiated to differentiated; from homogeneous and uniform to heterogeneous and multiform; and from an abundance and confusion of motion to a regimentation and loss of motion. These are mostly Spencer's own words. He does not speak of a passage from randomness to orderliness, or from more probable to less probable configurations of matter; but if these antitheses had been put to him, I feel sure he would have accepted them as fair descriptions of the trend or tendency of evolution.

Let us now study the principle of general evolution in its biological contexts to see if it actually works.

When used without further qualification, the word 'evolution' is generally taken to mean evolution in the genetic or Darwinian sense. Provided we confine ourselves to comparisons between grown-up organisms, evolution of this sort answers to Spencer's definition pretty well: there is indeed a passage from simple to complex and towards a differentiation and mutual dependence of parts. But Spencer's formulation completely fails to cope with the very real and important sense in which a frog's egg or embryo is more highly evolved than, say, a grown-up earthworm. Indeed, Spencer's conception of development (which I shall deal with in a moment) might lead us to think embryos *less* highly evolved than the adult forms of their own ancestors. This difficulty disappears if we take the view that evolution in the genetic sense is an evolution *of* developments – or, more exactly, an evolution of the genetic instructions that constitute the programme or specification of development. The genetic instructions that govern the development of a frog are much more elaborate and complicated than those that govern the development of an earthworm, and for that reason a frog's egg may be considered a more highly evolved object than an earthworm of any age.

But this leads to a paradox. Development itself is the golden example of an evolutionary process as Spencer conceived it to be. His thoughts constantly recurred to the evolution of tree from seed and of infant from 'germinal vesicle' (that is, egg). Unfortunately, development cannot be described as an evolution in the one essential sense in which genetic evolution *can* be so described. I have just said that we can get round the difficulty of being obliged to think an embryonic frog less highly evolved than an adult earthworm by thinking of an evolution of earthworm-making instructions into instructions for making frogs. Development is the carrying out of these instructions; it is a film that sticks faithfully to the book – an evolution, then, only in the sense of a translation, spelling out or 'mapping' of one kind of complexity into another kind of complexity. The adult is 'implicit' in the egg in the sense that one day it will be possible, after determining certain parameters, to read off the constitutional properties of the adult animal from a detailed knowledge of the chemical structure of the egg it arose from. Genetical evolution is entirely different; it is not a process of unfolding, and there is no useful sense in which the structure of a mammal can be said to be implicit in the structure of a protozoon.

I cannot make up my mind whether Spencer grasped this point or not. As early as 1852, in *The Development Hypothesis*, he wrote: 'The infant is so complex in structure that a cyclopedia is needed to describe its constituent parts. The germinal vesicle is so simple that it may be defined in a line.' Forty-five years of reflection must have confirmed him in this opinion, for the same sentences occur word for word in the first volume of the revised edition of *The Principles of Biology*. Yet in that same volume, grappling with the problem of how a peacock's tail comes to acquire its elaborate pattern, he made a rudimentary attempt to estimate what would now be called the amount of 'information' that must be present in a peafowl's egg to specify the pattern of one feather of the adult's tail. By erroneous reasoning[1] he came to the conclusion that 480,000 Weismannian 'determinants' would be required to specify the pattern of one feather alone. No wonder he declared that the 'organizing process transcends conception. It is not enough to say we cannot know it; we must say that we cannot even conceive it.' (To describe as 'inconceivable' what he himself could not conceive was one of Spencer's little weaknesses.)

However that may be, I think it must be clear that in describing both development and phylogenetic transformation as processes of 'evolution' we may be making a useful statement about one or about the other, but not, I fear, about both. They are altogether different phenomena. Biologists who use English as a scientific language *never* use the word 'evolution' to describe the processes of growth and development. (In France the usage is different.) They refrain because to do

[1] See *The Principles of Biology*, revised ed. (London, 1898), pp. 372–3. The computation is not possible even with the evidence now available to us; but if we were to attempt it we should certainly not assume that the individual elements of the pattern behaved as 'independent variables'.

so would be confusing and misleading. No such scruples weigh, alas, with biologists who speak about 'social evolution' or 'psychosocial evolution'. So far as I can make out from the writings of its various advocates, this superorganic evolution, as Spencer called it, has many manifestations, of which I shall mention only three and discuss only the third. They are (1) the Spencerian evolution of social organisation and of social institutions like governments, joint stock companies, banks and so on (this was what Spencer himself was mainly interested in). (2) What A. J. Lotka[1] called the 'exosomatic' (as opposed to ordinary or 'endosomatic') evolution of new sensory organs like spectacles, ear-trumpets, and ultraviolet spectrophotometers, or new motor organs like cutlery and guns. Spencer had some pointed and sensible remarks to make about evolution of this kind in his *Principles of Psychology*. These first two kinds of superorganic evolution are, of course, the consequences of a third: (3) cultural or psychosocial evolution, the secular accumulation of fact and fancy, knowledge and know-how, rules and rites, that is mediated through tradition.

By 'tradition' I mean 'the transfer of information through non-genetic channels from one generation to the next'.[2] I discussed psychosocial evolution at some length in the last of my Reith Lectures,[3] and tried to explain in just what sense psychosocial evolution represents a fundamentally new biological stratagem. But although psychosocial evolution is immensely important, it is also immensely obvious. Spencer did not ration himself austerely where explanations and examples were called for, but of the changes caused by the prodigious secular growth of the arts and sciences he says only 'the proposition is familiar and admitted by all. It is enough simply to point to this great phenomenon as one of the many forms of evolution we are tracing out.'[4] He spent much more time, unfortunately, in trying to demonstrate that the exercise of the mind had a direct hereditary effect on the capabilities of the brain in later generations. Perhaps this is why he believed in the inevitability of progress; for if his interpretation were true, social evolution would be cumulative and virtually irreversible.

[1] *Human Biology*, **17**, 167, 1945.

[2] *The Uniqueness of the Individual* (see p. 193 above, note 1), p. 141.

[3] *The Future of Man* (see p. 188 above, note 1).

[4] *The Principles of Psychology*, 4th ed. (London, 1899), vol. 1, § 158.

We know better: that we are all born into the Old Stone Age and in principle could stay there.

Psychosocial evolution differs from ordinary genetic evolution in three important ways: it is not mediated through genetic agencies; it is reversible, in the sense that what it has gained can in principle be wholly lost, and in one generation; and it is an evolution in the Lamarckian style, in the sense that a father's particular knowledge and skills and understanding can indeed be transmitted to his son, though not (as Spencer supposed) through genetic pathways. Common sense suggests that differences of this magnitude should be acknowledged by a distinction of terminology. The use of the word 'evolution' for psychosocial change is not a natural usage, but an artificial usage adopted by theorists with an axe to grind. If by any chance it *had* been a natural usage, people like myself on occasions like this would have said over and over again how wrong-headed it was, and how wise we should be to abandon it.

All who think about psychosocial 'evolution' agree that its inception marks a second great epoch of biological history. But I wonder: is it a second, or is it perhaps a third? The first must surely have been an evolution at the chemical or molecular level of integration – a process of which we can have no direct knowledge, for in a certain important sense all chemical evolution in living organisms stopped millions of years before even our faintest and most distant records of life began. So far as I know, no new *kind* of chemical compound has come into being over a period of evolution that began long before animals became differentiated from plants. Nor has there been any increase of chemical complexity; no chemically definable substance in any higher organism, for example, is more complex than a bacterial endotoxin. I have no views on the processes of evolution that brought new kinds of chemical compounds into existence, but I should not be surprised to find them very different from the forms of evolution that have been in progress since.

The point I wish to make is that evolution since those primordial days has been an evolution of structure at a higher level of integration than the chemical (using the word 'chemical' in the way it is used by chemists). It was these thoughts that led me to the discussion that now follows on the relationships between biological and thermodynamic order.

The sixteenth chapter – in effect the last – of the first edition of *First Principles* contains an argument on the phenomenon of equilibration from which Spencer ultimately drew 'a warrant for the belief, that Evolution can end only in the establishment of the greatest perfection and the most complete happiness'. This sentence cannot be found in the latest version of *First Principles*. Its place is taken by some sombre and, I must also say, rather confused reflections upon the ultimate state of the universe: for Spencer now enlarges and develops an argument of which, in the first edition, we see only the embryonic rudiments – an argument tending to the conclusion that unless something unforeseen and unforeseeable turns up, all things must 'beyond doubt' tend towards a universal quiescence, an 'omnipresent death'. What can have been responsible for the much greater weight he gave in his later thought to the phenomena of dissipation and dissolution?

The theory of general evolution was first hinted at in Spencer's *Development Hypothesis* of 1852. One year before, in the *Transactions* of the Royal Society of Edinburgh,[1] a Professor William Thomson (later Lord Kelvin) called attention to and elaborated upon some 'remarkable conclusions' arrived at by Clausius and Rankine after studying the properties of 'thermodynamic engines', engines that translate heat into mechanical work. The first law of thermodynamics (though not then so described) had already brought the reassuring news that heat, as a form of energy, could not be lost, for the total quantity of energy in the universe remained constant, no matter what its transformations. But though not lost to the universe, it now became certain that heat was 'irrevocably lost to man, and therefore "wasted", though not *annihilated*' in thermodynamic transactions; for (Thomson went on to explain) the conversion of heat into mechanical work depends on inequalities of temperature within the system, and these inequalities are progressively done away with in a great and universal process of levelling up.

Like the principle of general evolution, this second law of thermodynamics was in due course taken out of its native environment, here the pithead and the railway workshop, and generalised in much the same way as Spencer generalised the theory of evolution. With the growth of the science of statistical mechanics, it became possible to translate the second law into a statement about the history of a system of particles whose behaviour was known in the aggregate only, not individually. This historical statement declares that, in an isolated system, the pattern of the distribution of the elements within it passes from order towards randomness, from separatedness to mixedupness, and, in general, from less probable towards more probable configurations. Our sense of the fitness of things tells us that something is being lost in the process – orderliness, perhaps, or availability of energy, or thermodynamic competence – but the historical origins of the concept unhappily still persuade us to speak of a gain of something, of *entropy*, a quantity which, in its native context, is a simple ratio expressing the degree to which thermal energy is no longer available for the execution of mechanical work. '*Die Entropie strebt einem Maximum zu*' was Clausius's own formulation of the second law.

The most recent context for the general law of the decay of order and increase of entropy is in the theory of communication. A message encoded in symbols (for example, a Morse signal) owes its specificity, its property of being *this* message and not that message, to the particular configuration or sequence of the symbols, and a random or disorderly configuration of symbols does not make sense. The information capacity of a system of communication obviously depends on the range of different configurations of symbols at the command of the transmitting agent. In a sense, therefore, information capacity is a measure of order or, by a natural extension of the idea, of improbability; information capacity is thus analogous to negative entropy, and may be measured in formally similar terms. This formal similarity has led some people to declare that information *is* negative entropy, but the usage strikes me as perverse. There is much the same formal similarity between the equations for the diffusion of heat and for the diffusion of solutes, but (as I think Hogben somewhere remarked) we nowadays resist the temptation to refer to heat as a caloric fluid. 'Information', said Professor Norbert Wiener, in a passage more than usually full of negative entropy, 'is information, not matter or energy.' Elsewhere he points out that the concepts of information and of *pattern* are not coextensive: information is a concept normally (if not necessarily) applied to patterns which are spread out or must be read out in a series, in practice a time series.[1] Such is the case with the information in a gramophone or tape record and also, so it now appears, in a chromosome. The order matters.

By the end of the nineteenth century the philosopher could choose between alternative doctrines of world transformation, the one apparently contradicting the other. The principle of general evolution spoke of a secular increase of order, coherence, regularity, improbability etc., and Spencer's own derivation made it appear to follow logically from physical first principles; while the second law of thermodynamics, suitably generalised, spoke of a secular decay of order and dissipation of energy.

There can be no doubt that Spencer's thought took on a darker complexion in later years for essentially thermodynamic reasons; but such was the prestige of evolution theory that the second law of thermodynamics was over and over again described as a law of evolution, sometimes as *the* law of evolution. A. J. Lotka, the greatest of modern demographers, upheld this interpretation and applied it to biological evolution,[2] but to most people biological evolution and increase of entropy seem mutually contradictory ideas. 'Evolution', says Julian Huxley,[3] who can be spokesman for all who have thought likewise, 'is an anti-entropic process, running counter to the second law of thermodynamics with its degradation of energy and its tendency to uniformity'; and François Meyer[4] speaks of a principle of anti-chance at work among living things.

In fact the two concepts are not antithetical. That they are popularly supposed to contradict each other is due in part to a fairly obvious misunderstanding about the physical situations in which the two generalisa-

[1] *Cybernetics* (New York, 1948), p. 156, and *The Human Use of Human Beings* (London, 1950), p. 21.

[2] *The Elements of Physical Biology* (Baltimore, 1925). Lotka chose to regard evolution as the change undergone by the *total* system 'organisms + environment' conceived as an isolated system (or rather as a closed system with a known input of radiant energy). Conceived thus, the evolving system certainly obeys the second law, and there is much to be said for Lotka's viewpoint. But if we use the word 'evolution' to describe this general transformation, we shall have to invent another word to stand for evolution in its more usual biological sense.

[3] In his 'Introduction' to Teilhard de Chardin's *Phenomenon of Man*.

[4] *Problématique de l'évolution* (Paris, 1954).

tions hold good and make sense, and in part to a more subtle and correspondingly less obvious confusion of language.

Spencer himself, not unexpectedly, was unable to work the problem out; for this reason the final, revised edition of *First Principles* is much less satisfactory than the first, in which his thoughts were still untroubled by the ideas of universal mixing-up and running-down that grew out of the work of Boltzmann and Willard Gibbs. Spencer always believed that evolution had a natural limit, and came to an end when a certain state of equilibrium was reached, as he believed it always must be – for 'all terrestrial changes are incidents in the course of cosmic equilibration'. His arguments are not very clear because the equilibrium he refers to seems sometimes to mean a state of quiescence and rest, and at other times a steady state in which evolution and its exact opposite, dissolution, just cancel each other out. But dissolution eventually supervenes and the universe ends in a ruin of order. Evolution may start up again locally, and perhaps evolution and dissolution may alternate, but the matter must be left open because it is 'beyond the reach of human intelligence'. Spencer's answer seems to have reconciled the laws of evolution and of thermodynamics by supposing them to mark not incompatible but successive dynasties in a history of world order. But this will clearly not do, and we must seek other more convincing interpretations.

If we confine ourselves merely to considerations of energetics, there is no problem; or at least, there is no confusion. The second law of thermodynamics applies to *isolated* systems, systems in which there is no external trade in matter or energy; the law in no wise excludes the existence of sub-domains in which entropy may be decreasing,[1] though necessarily at the cost of a disproportionate increase of entropy elsewhere. The answer that has satisfied most physicists is that living organisms, thermodynamically *open systems*, are just such domains. They have been described as 'privileged domains', and this sounds well; though the corollary, that the rest of the system is underprivileged, sounds rather silly. But if we contemplate the overall

[1] For example, in a refrigerator, a heat engine in reverse in which heat flows from a colder to a warmer environment to increase the temperature of the latter. Needless to say the flow is far from spontaneous. See A. R. Ubbelohde, *Man and Energy* (Harmondsworth, 1963).

transformations of matter and energy within the biological system as a whole, as Lotka did, then of course nothing happens to contravene the second law.

A biologist might still protest that even if living organisms obey the letter of the second law of thermodynamics, they fail to observe it in the spirit. I should now therefore like to make a hasty and perhaps superficial attempt to clarify some of the real confusions of thought and language that underlie the entire argument. The confusions are genuine, even if my own interpretations turn out to be inexact.

I mentioned a little earlier the possibility of a grand abstract equivalence between, on the one hand, entropy, randomness, probability and nonsense; and on the other hand between thermodynamic order, 'order' as a biologist might use that word, improbability and information content. Each of these concepts or pairs of antitheses has been applied, sometimes critically, more often recklessly, to the description of living organisms. Let us therefore consider the sense, if any, (1) of regarding biological order as a form of thermodynamic order; (2) of describing organisms in evolution or in everyday life as seats or centres of improbability or anti-chance; and (3) of using the concepts and terminology of information theory to describe biological organisation. If these correspondences or equivalences should be found faulty then the antithesis between evolution and entropy, using both words in their widest senses, will disappear; for it will turn out that they refer to different properties pertaining to different physical situations.

1 First, then, thermodynamic order and biological order or structural organisation. So far as I can make out, all physicists who have considered the matter virtually equate the two, though Eddington[1] had some misgivings. Erwin Schrödinger, in his remarkable little book *What is Life?*,[2] says that living organisms maintain and add to their state of order by, in effect, feeding on negative entropy; by 'drinking orderliness from the environment' – or, to put it more temperately, by breaking down molecules coming from outside the system to pay the thermodynamic bill for synthesising molecules within the system. Schrödinger explicitly defines 'order' in the language of energetics, equating it to what was later to be called information capacity.

[1] *New Pathways in Science* (Cambridge, 1935).   [2] (Cambridge, 1944).

The shortcomings of an energetic definition of biological order seem to me to be these. In the first place, the thermodynamic cost of securing symmetry and order at the biological level of complexity – in macromolecules, for example – is trivial compared with the cost of synthesising the simpler molecules out of which they are constructed. The magnitude of entropy changes need therefore give one no significant information about the biological importance of the systems of ordering to which they refer. Secondly, the debit and credit columns of a thermodynamic transaction can be balanced by entries that do not bear on the distinctively biological properties of a macromolecular system. Two examples will illustrate this point.[1] The crystallisation of tobacco mosaic virus is accompanied by a slight *decrease* in the temperature of the solution from which it forms. A rise would have been expected, because the gain of order by the virus must be paid for by a gain of entropy elsewhere. How is this to be explained?

The explanation is that the hydrophobic surfaces of the virus subunits induce order in the water in which they move. When the units aggregate, the hydrophobic surfaces are in the interior of the virus, out of contact with water. The solvent thus becomes more disorganized, and this outweighs the order created in the virus. In this case the Second Law actually favours the creation of the well-ordered virus.

A cognate phenomenon is the reversible aggregation and dissociation of the subunits of certain 'cold-sensitive' enzymes, a matter of some importance for the control of metabolic processes within the cell. Pyruvic carboxylase is built of four similar subunits. On *cooling* to about 4° C the tetrad dissociates into subunits and enzymatic activity is lost. There must therefore be a compensating increase of orderliness in the ambient fluid.

It is phenomena of this kind that biologists have in mind when they suspect that biological systems obey the letter but not the spirit of the second law of thermodynamics. In general, the biologist's sense of the importance of order turns not on the distinction between order and disorder, but on the distinction between one form of ordering and

[1] These examples were brought to my attention by the late Dr Robin Valentine of the National Institute for Medical Research, and the explanation I quote is in his words. See also W. Kauzmann, *Advances in Protein Chemistry*, **24**, 37–47, 1959.

another; a chemically and thermodynamically trivial substitution of one nucleotide for another in a nucleic acid chain may make the difference between success and failure, or life and death. Willard Gibbs said that entropy was 'mixedupness'; biological order is not, or not merely, unmixedupness.[1]

2 Probability. Biologists in certain moods are apt to say that organisms are madly improbable objects or that evolution is a device for generating high degrees of improbability; I have already referred to M. Meyer's views on the workings of a principle of anti-chance. This is entirely in keeping with the idea that evolution generates negative entropy, because the second law of thermodynamics can be taken to declare that the spontaneous motion of all natural events is from less probable to more probable states.

I am uneasy about this entire train of thought for the following reason. Everyone will concede that in the games of whist or bridge any one particular hand is just as unlikely to turn up as any other. If I pick up and inspect a particular hand and then declare myself utterly amazed that such a hand should have been dealt to me, considering the fantastic odds against it, I shall be told by those who have steeped themselves in mathematical reasoning that its improbability cannot be measured retrospectively, but only against a prior expectation. Name a hand before the deal, I shall be told, and then everyone will be very much taken aback if it turns up.

For much the same reason it seems to me profitless to speak of natural selection's 'generating improbability'. When we speak, as Spencer was the first to do, of the survival of the fittest, we are being wise after the event: what is fit or not fit is so described on the basis of a retrospective judgement. It is silly to profess to be thunderstruck by the evolution of organism A if we should have been just as thunderstruck by a turn of events that had led to the evolution of B or C instead. The evolution of A was in fact the most probable net outcome of all the many selective forces that acted on its ancestors. The same goes for artificial selection. If I expose a culture of staphylococci to a certain concentration of penicillin, I *expect* (that is, I attach a high probability to) the evolution of a strain resistant to penicillin. Any

[1] See Dr Joseph Needham's searching and thoughtful analysis in his essay on 'Evolution and Thermodynamics' in *Time: The Refreshing River* (London, 1943).

other outcome would be improbable and would require a special explanation.

It is not only in their evolution but in their growth and persistence from day to day that organisms have been said to be wildly improbable phenomena. By this is meant, I suppose, that if all the ingredients of the world or the solar system were to be shaken up in a dice box of divine dimensions, the emergence from it of a configuration like an earthworm would be most improbable. How very true! But in the physical circumstances that actually prevail, the growth and maintenance of an organism is the most probable outcome of the events it is taking part in. When I eat a meal, for example, I expect part of it to turn into more of myself. In particular, I shall expect some of it to turn into the chemical substance characteristic of my own blood group, group B. So very highly probable is it that substance B will be formed that a court of law will no longer countenance the possibility of my manufacturing a certain amount of blood group substance A – though substances A and B are almost identical physically and chemically.

Similarly, if I spark a mixture of gaseous hydrogen and oxygen, the most probable consequence is the formation of water, though water is a more highly 'organised' substance, as the biologist uses that word, than the elements out of which it was compounded. What *may* be said, I think, is that the conjunction of circumstances of which the formation of water is the most probable outcome is itself very improbable – I mean, the coming together of gaseous oxygen and hydrogen and the application of a spark to the mixture. Certainly organisms, to remain alive, generate improbable, conspicuously non-random situations, and pay a heavy price in energy for doing so; but that is hardly an arresting thought.

3 Finally, 'information'. The ideas and terminology of information theory would not have caught on as they have done unless they were serving some very useful purpose. It seems to me that they are highly appropriate in their proper context, where we have to do with storing information or sending messages; chromosomes, for example, convey from one generation to the next a message about how development is to proceed, and in speaking of genes and chromosomes the language of information theory is often extremely apt. But I feel we have to be on our guard against treating information content as a

measure of biological organisation. If it were indeed so, then, as Waddington has pointed out,[1] we should be obliged to infer that complexity or degree of organisation increased very little in biological development. For development, at all events when it occurs in a nearly closed system, is, from the standpoint of information theory, merely a verbose and repetitious spelling out or biological rewording of the information encoded in the chromosomes. This is a most unhelpful description of development.

In my opinion the audacious attempt to reveal the formal equivalence of the ideas of biological organisation and of thermodynamic order, non-randomness and information must be judged to have failed. We still seek a theory of order in its most interesting and important form, that which is represented by the complex functional and structural integration of living organisms. For the present we must be content to say that, in biology, the concepts of entropy and thermodynamic order are appropriate when we are dealing with matters of energetics; of information theory when we are studying how messages are sent and acted upon; and of probability where we are dealing with phenomena that have a random element, as in predicting the outcome of breeding experiments. In our moods of abstract theorising we tend to forget how great and how diverse are the functional commitments of biological macromolecules. They insulate, they fill out; they fetch and carry; they prevent the organism as a whole from falling apart or from dissolving in water; they prop up, they protect; they attack and defend; they store energy and catalyse its transfer; they store information and convey messages, and sometimes they themselves *are* messages. The successful prosecution of all these activities depends upon properties more complex, various and particular than can be written down in the language of energetics or information theory.

Where a consortium of brilliant and imaginative theorists has failed, so far, to provide us with the right theoretical equipment for studying biological organisation, we need not wonder that Herbert Spencer, working pretty well single-handed, failed too. His system of general

---

[1] 'Architecture and Information in Cellular Differentiation', in J. A. Ramsay and V. B. Wigglesworth (eds), *The Cell and the Organism*. For discussions of information theory in biology, see H. Quastler (ed.), *Essays on the Use of Information Theory in Biology* (Illinois, 1953); W. M. Elsasser, *The Physical Foundation of Biology* (Oxford, 1958).

evolution does not really work; the evolution of society and of the solar system are different phenomena, and the one teaches us next to nothing about the other. Development on the one hand, and the secular transformations of species on the other hand, are indeed both evolutionary processes, but in senses so different that to describe one as an 'evolution' makes it imperative to find some different word to describe the other. But for all that, I for one can still see Spencer's system as a great adventure, and now that I know my way about those thick, square volumes I do not feel I am taking leave of them for good.

D'ARCY THOMPSON AND
GROWTH AND FORM

D'Arcy Wentworth Thompson was an aristocrat of learning whose intellectual endowments are not likely ever again to be combined within one man. He was a classicist of sufficient distinction to have become President of the Classical Associations of England and Wales and of Scotland; a mathematician good enough to have had an entirely mathematical paper accepted for publication by the Royal Society; and a naturalist who held important chairs for sixty-four years, that is, for all but the length of time into which we must nowadays squeeze the whole of our lives from birth until professional retirement. He was a famous conversationalist and lecturer (the two are often thought to go together, but seldom do), and the author of a work which, considered as literature, is the equal of anything of Pater's or Logan Pearsall Smith's in its complete mastery of the bel canto style. Add to all this that he was over six feet tall, with the build and carriage of a Viking and with the pride of bearing that comes from good looks known to be possessed.

D'Arcy Thompson (he was always called that, or D'Arcy) had not merely the makings but the actual accomplishments of three scholars. All three were eminent, even if, judged by the standards which he himself would have applied to them, none could strictly be called great. If the three scholars had merely been added together in D'Arcy Thompson, each working independently of the others, then I think we should find it hard to repudiate the idea that he was an amateur, though a patrician among amateurs; we should say, perhaps, that great as were his accomplishments, he lacked that deep sense of engagement that marks the professional scholar of the present day. But they were not merely added together; they were integrally – Clifford Dobell said chemically – combined. I am trying to say that he was not one of those who have made two or more separate and somewhat incongruous reputations, like a composer-chemist or politician-novelist, or like the one man who has both ridden in the Grand National and become a Fellow of the Royal Society, the world's oldest and most famous society of scientists; but that he was a man who comprehended many things with an undivided mind. In the range and quality of his learning, the uses to which he put it, and the style in which he made it known I see not an amateur, but, in the proper sense of that term, a natural philosopher – though one dare not call him so without a hurried qualification, for fear he might be thought to have practised what the Germans call *Naturphilosophie*.

Let me now try to describe the environment in which D'Arcy the scientist lived and worked. When D'Arcy flourished, British zoology, after fifty years, was still almost wholly occupied with problems of phylogeny and comparative anatomy, that is, with the apportioning out of evolutionary priorities and the unravelling of relationships of descent. Comparative anatomy has many brilliant discoveries to its credit; for example, the discovery that the small bones of the middle ear – those which transmit vibrations from the ear drum to the organ of hearing – are cognate with bones which in the remote ancestors of mammals had formed part of the hinges and articulations of the lower jaw; that the limbs of terrestrial vertebrates had evolved along a just discernible pathway from the paired fins of fish; that the muscles which move the eyeballs derive in evolution from the anterior elements of a segmental musculature which at one time occupied the body from end to end. But although later work refined upon them or corrected them here or there, all these discoveries had been made in the nineteenth century. When D'Arcy took his chair, the great theorems of comparative anatomy had already been propounded, and nearly all the great dynasties in the evolutionary history of animals were already known. By 1917, the date of the first edition of D'Arcy's essay *On Growth and Form*, British zoology was far gone in that decline from which a small group of 'comparative physiologists' was to rescue it in the middle '20s – in due course (it has been rudely said) to impose upon it a hegemony of their own. The work of phylogeny and comparative anatomy is not yet all done. We are still uncertain about the affinities of whales, though we may be quite sure that purely anatomical research will not reveal them; the comparative anatomy of the lymphatic system has hardly been attempted; and there is doubtless much to be done among the parish registers of evolution, in the attempt to trace lines of descent within families of animals, or even within genera. But comparative anatomy is no longer thought of as the central discipline of zoology; D'Arcy Thompson was the first man in this country to challenge its pretensions and to repudiate the idea that zoological learning consisted of so many glosses on an evolutionary text.

In D'Arcy Thompson's earliest writings there is little to suggest that he would one day slough off the coils of evolutionary anatomism, though one of his papers – 'On the Nature and Action of Certain Ligaments' (1884) – is evidence that he was interested in bones for how they worked rather than for what they might have to say about their owners' evolutionary credentials. Later, writing again of ligaments, he said:

The 'skeleton', as we see it in a Museum, is a poor and even a misleading picture of mechanical efficiency. From the engineer's point of view, it is a diagram showing all the compression-lines, but by no means all of the tension-lines of the construction . . . it falls to pieces unless we clamp it together, as best we can, in a more or less clumsy and immobilized way. In preparing or 'macerating' a skeleton, the naturalist nowadays carries on the process till nothing is left but the whitened bones. But the old anatomists . . . were wont to macerate by easy stages; and in many of their most instructive preparations, the ligaments were intentionally left in connection with the bones, and as part of the 'skeleton'.

Whitened immobile bones, rearticulated with bits of wire, were indeed symbolic of the evolutionary anatomism which had 'all but filled men's minds during the last couple of generations'.

The treatment of bones and other bodily structures as so many archives of evolution angered D'Arcy for two chief reasons; first, because his contemporaries and immediate predecessors (it is difficult to speak of the contemporaries of a man who lived so long, but here and elsewhere I centre his life about the year of the publication of *Growth and Form*) had no real curiosity beyond the evolutionary pedigree of an organism or an organ: any enquiry into the action of contemporary physical causes seemed to them to belong to some other science; and secondly because the comparative anatomists, in spite of their devotion to the study of its consequences, were no more than idly curious about the *mechanism* of evolution; they accepted the contemporary and far from adequate form of Darwinism in much the same way as nicely brought up people often accept their religion, that is, in a manner that contrives to be both tenacious and perfunctory. D'Arcy's own opinion was that we should look to the action of contemporary and immediately impingent physical causes for the explanation of an animal's growth and form. What causes the spicules of sponges to take their characteristic shapes? D'Arcy sought the answer in the phenomena of crystallisation and of adsorption and diffusion, instead of being content with the explanation for which he makes Minchin spokesman, viz. that 'The forms of the spicules are the result of adaptation to the requirements of the sponge as a whole, produced by the action of natural selection upon variation in every direction', and that their regular form is 'a phylogenetic adaptation, which has become fixed and handed on by heredity, appearing in the ontogeny as a prophetic adaptation'. For sponges and spicules one could substitute other organisms and other organs: the formula would accommodate all comers. Then again, D'Arcy tried (very imperfectly, to be sure) to envisage how the physical forces acting upon them might have shaped the shells of the Foraminifera, instead of being content to see in their diverse patterns no more than the branches of a hypothetical family tree. In embryology, evolutionary anatomism seemed particularly inexcusable, for real embryos, unlike hypothetical ancestors, are tangible present objects, and amenable therefore to 'causal' investigation in the sense in which the physicist employs that term. Yet Hertwig declared for *post hoc, propter hoc*, holding that the chronological anatomy of embryos provided causes sufficient to explain development, and Balfour valued embryology mainly for its testimony of descent. Of course, it was not D'Arcy himself, but long before him Roux and His who founded modern analytic embryology by trying to introduce a little dynamics into evolutionary dynastics; but for so long did the spirit of anatomism prevail that even when I was a student at Oxford, the causal analysis of development was separated from descriptive embryology and treated as a thing apart. No wonder D'Arcy was an anti-Darwinian! Believing as he did that present phenomena should be explained by present causes, he saw the appeal to deep historical antecedents as an evasion of responsibility – all the more culpable for being made with the authority of what was, at the time, a most imperfect evolutionary theory.

Here I believe that D'Arcy was as much in error as those whose doctrines he endeavoured to correct. I must make my point at length, because it is the substance of the charge that D'Arcy was somehow 'unbiological' in his way of thinking, and it explains why, although he was surely right to be annoyed with his more austerely anatomical colleagues, they in turn were not wholly to blame for feeling annoyed with him.

Consider the argument set out in chapter XVI, 'On Form and Mechanical Efficiency', in the essay *On Growth and Form*. Here D'Arcy tells us, amongst other things, of the fitness to their several purposes of bone and bones. The shafts of the humerus and femur are hollow cylinders of a dense compact bone which is thicker in the middle than at either end, for it is sound engineering to have the thickness vary with the bending moments. At each end the shafts widen, and the compact bone thins out, enclosing within it a thick layer of bone of a more open texture – cancellar bone, made of intersecting laminae or trabeculae of bony matter. In a section cut lengthwise through the head of, for example, the human femur, one can see that the bony trabeculae are not arranged haphazardly, but that they form two systems of curves, intersecting roughly at right angles, which are (to a fair approximation) a structural embodiment of the system of stresses to which the bone is exposed in normal life. (It is an old story this, which goes back to Julius Wolff and Herman Meyer.) The entire arrangement clearly represents an adaptation; what kind of adaptation could it be?

The trabeculae, D'Arcy reminds us, are not permanent structures: they are constantly being broken down and formed anew, and if by mischance a bone should be broken and should reunite in some abnormal fashion, the trabeculae will shape themselves into a new pattern governed by the new and altered system of stresses and strains. This, then, should give us almost direct evidence of what must happen in ordinary development: the trabeculae begin by being 'laid down fortuitously in any direction within the substance of the bone', but end in that functionally apt pattern which seems so clearly to represent the engravery of actual use. 'Here, for once, it is safe to say that "heredity" need not and cannot be invoked to account for the configuration and arrangement of the trabeculae: for we can see them, at any time of life, in the making, under the direct action and control of the forces to which the system is exposed.' D'Arcy could have drawn the same conclusion from a study of the forms of joints, for when a long bone is broken and the broken ends fail to reunite, they sometimes become hinged to each other in what is anatomically an almost perfect joint. What better evidence could there be that joints too, with all the niceties of their patterns of articulation, are shaped by use?

Yet they are not so. For all their fitness to mechanical purposes, the patterns of bone and bones are not, in the first instance, moulded by the demands of use; the evidence of remodelling and regeneration shows that they *could* be so, and that under special circumstances they are so; but bones will develop in an anatomically almost perfect fashion even when deprived of innervation or transplanted into positions where they can neither move nor be moved. To explain the shapes of bones we must look elsewhere than to the mechanical forces that act in an individual's own lifetime. We need not doubt that D'Arcy's forces acted once upon a time in directing the pathway of evolution, but if that is so, then the problem is just what D'Arcy supposed it not to be: a matter of history. The whole point is that the forces which did at one time shape limbs, or set the limits within which the shapes of limbs must fall, were translated into those developmental *instructions* about limb-making that now form part of our genetic heritage; the problem of the development of limbs is, first, to break the chemical code which embodies the instructions, and second, to find out how the instructions take effect. Here, too, lies the sense of that venerable old antithesis between 'preformation' and 'epigenesis': the instructions are ready-made, but their fulfilment is epigenetic; heredity proposes and development disposes. No one denies the essential truth of what D'Arcy Thompson had to say about the influence of physical and mechanical forces; he simply mistook their context. Perhaps he was aware of having done so when, on a later page, he admits that 'a principle of heredity' may have much to do with the matter. But he never quite realised that he and the comparative anatomists were giving not rival answers to the same question but different answers to two different questions. D'Arcy's answer is to the question: What physical agencies formed the basis of natural selection, and so caused one particular set of instructions about limb-formation to prevail? So construed, all that he has to say is relevant. After all, the elementary forces whose action he contemplated were no different in any yesterday from what they are at present: 'a snow crystal is the same today as when the first snows fell'.

Because he spoke out impatiently against contemporary orthodoxy, D'Arcy Thompson is often thought of as a great innovator; but the angle subtended between an innovator and his contemporaries gets smaller in the more distant view, and at a distance of forty years it no longer seems to us that he stood so far to one side of his anatomical colleagues. Anyone who reads *Growth and Form* attentively will soon discern that D'Arcy was an anatomist himself; the life that appears in his pages is usually still life, and it was product rather than performance that usually engaged his attention. I am well aware that his conception of form was essentially dynamical; but though he did indeed declare that any given shape was to be thought of in terms of some orderly array of inequalities of growth-rates, yet in his 'Method of Transformations' (which I shall mention later) he did in fact compare the final products of two processes of transformation and not the processes themselves.

If D'Arcy was an anatomist, he was the first completely modern anatomist, in that his conception of structure was of molecular as well as of merely visible dimensions, his thought travelling without impediment across the dozen orders of magnitude that separate the two. The advances that have occurred in modern biophysics and structural biochemistry are comprehended within D'Arcy's way of thinking. We all now understand, though the idea was revolutionary in its day, that molecules themselves have shapes as well as sizes: some are long and thin, others broad and flat, some straight, some branched. We also know that crystalline structure is enjoyed by the huge molecules of proteins, nucleic acids, and polysaccharides as well as by the 'crystalloids' of an older terminology, and that the structure of the cell surface and of some of the 'organelles' enclosed by it must be interpreted in molecular terms. D'Arcy would have delighted in the modern X-ray crystallography of 'biological' compounds and the penetrating insight of the higher powers of the electron microscope; particularly would he have rejoiced in the modern solution of the structure of deoxyribonucleic acid – a brilliant feat of chemical anatomy that provides us for the first time with some physical conception of the 'gene'. Biology, and the chemistry and physics that go with it, have grown more rather than less anatomical in recent years, and the anatomy is indeed Thompsonian, one which recognises no frontier between biological and chemical form.

This is the right moment to explain D'Arcy's own 'philosophy' of living organisms. He believed (a) that the laws of the physical sciences apply to living organisms, and (b) that living organisms do nothing to contravene those laws. These propositions are sometimes taken to import that biology is, or soon will be, nothing more than a kind of super physics-and-chemistry. In reality they do nothing of the kind. Biology deals with notions that are contextually peculiar to itself – with heredity, development and sexuality; with reflex action, memory and learning; with resistance to disease and disease itself. These things are no more part of physics and chemistry than are the Bank Rate or the British Constitution. We are mistaking the direction of the flow of thought when we speak of 'analysing' or 'reducing' a biological phenomenon to physics and chemistry. What we endeavour to do is the very opposite: to assemble, integrate or piece together our conception of the phenomenon from our particular knowledge of its constituent parts. It was D'Arcy's belief, as it is also the belief of almost every reputable modern biologist, that this act of integration is in fact possible. He stopped short of supposing that the act of integration would eventually irrupt upon matters of the spirit: 'Of how it is that the soul informs the body, physical science teaches me nothing . . . nor do I ask of physics how goodness shines in one man's face, and evil betrays itself in another.' But D'Arcy makes no other mention of these matters; nor shall we.

The essay On Growth and Form (so he described it, though it is nearly 800 pages long) was D'Arcy's magnificent attempt to put his philosophic principles to work. The attempt was successful in so far as it depended upon his geometrical insight, and courageous (though inevitably often faulty) whenever the physicist got the upper hand. The biologist in him, strangely enough, was the weakest member of the team. D'Arcy's treatment of form, then, is generally illuminating, particularly when he tells us how pervasive are certain elementary forms – the unduloid and catenoid, as well as their simpler relatives the sphere and cylinder, the logarithmic spirals of the horns and shells that grow by accretion while remaining unchanged in shape, the geodetic lines of the thickening of plant cell walls, the geometric packing adopted by cellular aggregates. When it came to the physicist's turn, and the attempt to explain the shapes of cells or of spicules, or the mechanisms of amoeboid movement or phagocytosis, then indeed his simple armament of surface tension, viscosity, diffusion and adsorption was not powerful enough by far. We cannot reproach D'Arcy for having failed to solve problems which, most of them, defy us yet; but it is a fair comment that D'Arcy himself made no move to solve them, whether by doing experiments himself or by suggesting experiments that might have been done by others.

D'Arcy Thompson was sometimes accused of being too much the geometer in his way of thinking, because of his determination to see simple regularity where a literal-minded person would say it did not exist: the spheres he saw were not quite spherical, the polygons not quite regular, the transformations not quite orthogonal, the bony trabeculae an inaccurate representation of stress and strain. It is an old and wearisome story, and in finding this new context for it I believe that D'Arcy's critics were completely wrong. Surely we must always begin by seeking regularities. There is *something* in the fact that the atomic weights of the elements are so very nearly whole numbers; that there is a certain periodicity in the properties of the elements when they are written down in the order of their ascending weights; that hereditary factors tend to be inherited not singly but in groups equal in number to the number of pairs of chromosomes. Unless we see these regularities and strive to account for them, we shall not equip ourselves to understand or even perhaps to recognise the existence of isotopes, the significance of atomic number, or the phenomenon in genetics of crossing over. We act either as D'Arcy does, simplifying and generalising until the facts confute us, or, like those in whom Bacon saw a chief cause of the retardation of learning, we must betake ourselves to 'complaints upon the subtleties of nature, the secret recesses of truth, the obscurity of things, and the infirmity of man's discerning power'.

D'Arcy must be acquitted of the charge of striving without just reason for simplicity, but the charge that the biologist in him was strangely unperceptive cannot so easily be dismissed. Here is an example taken from *Growth and Form* in which even a layman should be able to see a certain perversity of reasoning. It comes from chapter II, 'The Rate of Growth'.

To us nowadays, as fifty years ago to the Bostonian anatomist Charles Minot, it seems obvious that the *norm* of biological growth – the standard to which all actual instances of growth must be referred – is growth by compound interest, sometimes called logarithmic or exponential growth. A house grows by accretion: the bricks which represent the unit increments of its growth stay put and do not grow themselves. But the central characteristic of biological growth is that that which is formed by growth is itself capable of growing: the interest earned by growth becomes capital and thereupon earns interest on its own behalf. We must therefore plot the growth of organisms not against an ordinary arithmetic scale, equal subdivisions of which represent equal additions of size, but against a logarithmic scale, of which equal subdivisions represent equal *multiples* of size. The simplest case is when the interest earns interest at the same rate as the capital to which it is added; in such a case, the logarithm of size will give a straight line when plotted against age. D'Arcy missed the point of Minot's appeal that this was the form in which curves of biological growth should be represented, just as, later, he was to miss the point of J. S. Huxley's analogous treatment of the phenomenon of differential growth, in which the multiplication-rate of one part of the body is measured against that of another. D'Arcy preferred to deal with the arithmetical or 'simple interest' type of growth curve, with its first derivative, growth-rate, and its second derivative, acceleration of growth. This treatment not only hid from him much that was important, but led him to attach importance to things of no great weight – for example, the time of life at which the arithmetic growth-rate is at a maximum, seen as the point of inflexion of the integral curve of growth.

The best and most famous chapter of *Growth and Form* is that which embodies D'Arcy Thompson's own most completely original contribution to biology: the 'Method of Transformations' in the chapter 'On the Comparison of Related Forms'. Consider the shapes of two organisms of related genera. Being closely related, the two shapes will be 'homoeomorphic' – that is, roughly speaking, they will be qualitatively the same, so that the one could be changed into the other, as two different faces could be, by some process of plastic remodelling. But although two such shapes may be qualitatively similar, they will in fact differ in a multitude of particular little ways. The anatomist's method of comparing shapes is to do so piecemeal, feature by feature, with perhaps an occasional measurement of proportion. D'Arcy saw that all these particular little differences of disposition, angle, length and ratio might be simply the topical expressions of some one comprehensive and pervasive change of shape. He grasped the transformation *as a whole*. So it might be, for example, with the change of shape produced by distorting a sheet of rubber on which has been drawn a house, or face, or any other figure whatsoever: the shapes of the drawings change in every single particular, but the transformation as a whole might be defined by some quite simple formula describing the way the rubber had been stretched. So also with the transformation that may be produced by tilting a lantern screen away from its normal perpendicular position; whatever is shown on the screen will be transformed in every detail, but in a manner which can be summarily defined by formulae expressing the direction and the degree of tilt. In all such cases the best way of grasping the transformation is to take, as its subject, some simple and regular figure; for mathematical purposes one takes the ordinary 'Cartesian grid' of graph paper, an orthogonal system of equally spaced straight lines. This was the way in which D'Arcy expressed the relationship between *Diodon* and *Orthagoriscus* or between *Argyropelecus* and *Sternoptyx* (see diagrams), to choose two among the best of a number of good examples. The grids superimposed upon these figures give one instantly the sense and trend of the transformations; we may think, if we like, that *Orthagoriscus* is a *Diodon* living in some quite remarkably non-Euclidian principality of the ocean, or that *Diodon* is an *Orthagoriscus* of ordinary Cartesian seas. The lesson to be learnt is that we do not have to seek a hundred different explanations of the hundred particular differences between the one form and the other; *one* system of 'morphogenetic forces' may perhaps account for all the differences between the two.

It is hardly necessary to say that the whole treatment is an oversimplification, and that the transformations figured by D'Arcy could not conceivably have happened in real life. In real life, one adult does not change into another adult, but two related embryos turn into two related adults. However, D'Arcy's somewhat elliptical treatment does not make his one important lesson any the less clear. The reason why D'Arcy's method has been so little used in practice (only I and one or two others have tried to develop it at all) is because it is analytically

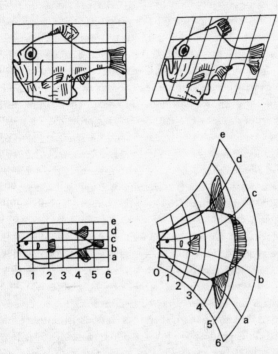

Upper left, *Argyropelecus*; upper right, *Sternoptyx*; lower left, *Diodon*; lower right, *Orthagoriscus*. From D'Arcy Thompson, *Growth and Form* (Cambridge University Press).

unwieldy. The methods later developed by J. S. Huxley, though far less comprehensive and ambitious, were much more usable and informative, and they were widely taken up from the moment they were first described. D'Arcy Thompson gave rather perfunctory praise to these developments, believing Huxley's methods of analysis to be implicit in his own. It is in the 'new edition' of *On Growth and Form*, of 1942, that D'Arcy hints at this opinion; I have been writing all the time of the edition that came out at the corresponding period of the First World War. D'Arcy's reputation as a scientist rests almost wholly upon his 'old' edition, and some of D'Arcy's colleagues thought him unwise in his attempts to bring it up to date.

There is one more thing to be said of *Growth and Form*, and perhaps it is the most important thing of all; it relates to an accomplishment in which D'Arcy himself took the utmost pride, and in which he knew in his heart he had no equal. I think that *Growth and Form* is beyond comparison the finest work of literature in all the annals of science that have been recorded in the English tongue. There is a combination here of elegance of style with perfect, absolutely unfailing clarity that has never to my knowledge been surpassed. To be sure, much of D'Arcy's writing sounds old-fashioned – his prose has a longer stride than we can keep up with nowadays – and the scholarly allusions and digressions and the little graces of writing may be somewhat overdone; but this and all the other decorative matter is simply the *fioritura* of the perfectly accomplished singer. *Growth and Form* will remain for ever worth reading as a text in the exacting discipline of putting conceptions accurately into words.

The influence of *Growth and Form* in this country and in America has been very great, but it has been intangible and indirect. It is to be seen in anyone who, having read it, tries to write a little more clearly and with at least an attempt at grace; or who realises as he turns its pages, perhaps for the first time, that science cannot be divided into what is up to date and what is merely of antiquarian interest, but is to be regarded as the product of a growth of thought. Most clearly of all it is to be seen in the complete matter-of-factness with which we now accept certain beliefs that D'Arcy, as a natural historian, had to fight for: not merely that the physical sciences and mathematics offer us the only pathway that leads to an understanding of animate nature, but also that the true beauty of nature will be revealed only when that understanding has been achieved. To us nowadays it seems obvious that the picture we form in our minds of nature will be the more beautiful for being brightly lit. To many of D'Arcy's contemporaries it must have seemed strange and even perverse that he should have combined a physico-mathematical analysis of nature with, at all times, a most intense consciousness of its wonder and beauty; for at that time there still persisted the superstition that what is beautiful and moving in nature is its mystery and its unrevealed designs. D'Arcy did away for all time with this Gothic nonsense: a clear bright light shines about the pages of *Growth and Form*, a most resolute determination to unmake mysteries.

It is by such diffused and widely pervasive effects as these that we must measure the influence of *Growth and Form* upon biological science. Of direct influence that can be traced in pedigrees of teaching or research there is little. In a generation's time there will be no one alive who heard D'Arcy's lectures, and no one to declare from personal knowledge that he knew the animal kingdom inside out. D'Arcy had only one pupil of more than ordinary distinction, and *he* made his name in descriptive zoology of a most un-Thompsonian kind. D'Arcy founded no school, as Sherrington did, so that no lineage of research can be traced back directly to sources in his mind. But then, he did no research in the modern sense; he was, as I said to begin with, a natural philosopher, one who by reflection rather than by intervention or experiment arrived at a certain imperfect but nevertheless whole conception of that science in which God has been slowest to reveal Himself a geometer. It was a conception expressed for the most part in a modern scientific idiom, and with a beauty and clarity of writing that may never be surpassed.

THE PHENOMENON OF MAN

Everything does not happen continuously at any one moment in the universe. Neither does everything happen everywhere in it.

There are no summits without abysses.

When the end of the world is mentioned, the idea that leaps into our minds is always one of catastrophe.

Life was born and propagates itself on the earth as a solitary pulsation.

In the last analysis the best guarantee that a thing should happen is that it appears to us as vitally necessary.

This little bouquet of aphorisms, each one thought sufficiently important by its author to deserve a paragraph to itself, is taken from Père Teilhard's *The Phenomenon of Man*.[1] It is a book widely held to be of the utmost profundity and significance; it created something like a sensation upon its publication in France, and some reviewers hereabouts called it the Book of the Year – one, the Book of the Century. Yet the greater part of it, I shall show, is nonsense, tricked out with a variety of metaphysical conceits, and its author can be excused of dishonesty only on the grounds that before deceiving others he has taken great pains to deceive himself. *The Phenomenon of Man* cannot be read without a feeling of suffocation, a gasping and flailing around for sense. There is an argument in it, to be sure – a feeble argument, abominably expressed – and this I shall expound in due course; but consider first the style, because it is the style that creates the illusion of content, and which is a cause as well as merely a symptom of Teilhard's alarming apocalyptic seizures.

The Phenomenon of Man stands square in the tradition of *Naturphilosophie*, a philosophical indoor pastime of German origin which does not seem even by accident (though there is a great deal of it) to have contributed anything of permanent value to the storehouse of human thought. French is not a language that lends itself naturally to the opaque and ponderous idiom of nature-philosophy, and Teilhard

[1] (London, 1959).

has accordingly resorted to the use of that tipsy, euphoristic prose-poetry which is one of the more tiresome manifestations of the French spirit. It is of the nature of reproduction that progeny should out-number parents, and of Mendelian heredity that the inborn endow-ments of the parents should be variously recombined and reassorted among their offspring, so enlarging the population's candidature for evolutionary change. Teilhard puts the matter thus: it is one of his more lucid passages, and Mr Wall's translation, here as almost every-where else, captures the spirit and sense of the original.

Reproduction doubles the mother cell. Thus, by a mechanism which is the inverse of chemical disintegration, *it multiplies without crumbling*. At the same time, however, it transforms what was only intended to be prolonged. Closed in on itself, the living element reaches more or less quickly a state of immo-bility. It becomes stuck and coagulated in its evolution. Then by the act of reproduction it regains the faculty for inner re-adjustment and consequently takes on a new appearance and direction. The process is one of pluralization in form as well as in number. The elemental ripple of life that emerges from each individual unit does not spread outwards in a monotonous circle formed of individual units exactly like itself. It is diffracted and becomes iridescent, with an indefinite scale of variegated tonalities. The living unit is a centre of irresistible multiplication, and *ipso facto* an equally irresistible focus of diversification.

In no sense other than an utterly trivial one is reproduction the inverse of chemical disintegration. It is a misunderstanding of genetics to suppose that reproduction is only 'intended' to make facsimiles, for parasexual processes of genetical exchange are to be found in the simplest living things. There seems to be some confusion between the versatility of a population and the adaptability of an individual. But errors of fact or judgement of this kind are to be found throughout, and are not my immediate concern; notice instead the use of adjectives of excess (misuse, rather, for genetic diversity is not indefinite nor multiplication irresistible). Teilhard is for ever shouting at us: things or affairs are, in alphabetical order, astounding, colossal, endless, enor-mous, fantastic, giddy, hyper-, immense, implacable, indefinite, inexhaustible, inextricable, infinite, infinitesimal, innumerable, irresist-ible, measureless, mega-, monstrous, mysterious, prodigious, relent-less, super-, ultra-, unbelievable, unbridled or unparalleled. When something is described as merely *huge* we feel let down. After this softening-up process we are ready to take delivery of the neologisms: biota, noosphere, hominisation, complexification. There is much else in the literary idiom of nature-philosophy: *nothing-buttery*, for example, always part of the minor symptomatology of the bogus. 'Love in all its subtleties is nothing more, and nothing less, than the more or less direct trace marked on the heart of the element by the psychical convergence of the universe upon itself.' 'Man discovers that he is *nothing else than evolution become conscious of itself*,' and evolution is 'nothing else than the continual growth of . . . "psychic" or "radial" energy'. Again, 'the Christogenesis of St Paul and St John is nothing else and nothing less than the extension . . . of that noogenesis in which cosmogenesis . . . culminates'. It would have been a great disappointment to me if Vibration did not somewhere make itself felt, for all scientistic mystics either vibrate in person or find themselves resonant with cosmic vibrations; but I am happy to say that on page 266 Teilhard will be found to do so.

These are trivialities, revealing though they are, and perhaps I make too much of them. The evolutionary origins of consciousness are indeed distant and obscure, and perhaps so trite a thought does need this kind of dressing to make it palatable: 'refracted rearwards along the course of evolution, consciousness displays itself qualitatively as a spectrum of shifting hints whose lower terms are lost in the night' (the roman type is mine). What is much more serious is the fact that Teilhard habitually and systematically cheats with words. His work, he has assured us, is to be read, not as a metaphysical system, but 'purely and simply as a scientific treatise' executed with 'remorseless' or 'inescapable' logic; yet he uses in metaphor words like energy, tension, force, impetus and dimension *as if* they retained the weight and thrust of their special scientific usages. Consciousness, for example, is a matter upon which Teilhard has been said to have illuminating views. For the most part consciousness is treated as a manifestation of energy, though this does not help us very much because the word 'energy' is itself debauched; but elsewhere we learn that consciousness is a dimension, or is something with mass, or is something corpuscular and particulate which can exist in various degrees of concentration, being sometimes infinitely diffuse. In his lay capacity Teilhard, a naturalist, practised a comparatively humble and unexacting kind of science, but he must have known better than to play such tricks as these. On page 60 we read:

The simplest form of protoplasm is already a substance of unheard-of complexity. This complexity increases in geometrical progression as we pass from the protozoon higher and higher up the scale of the metazoa. And so it is for the whole of the remainder always and everywhere.

Later we are told that the '*nascent* cellular world shows itself to be already infinitely complex'. This seems to leave little room for improvement. In any event complexity (a subject on which Teilhard has a great deal to say) is not measurable in those scalar quantities to which the concept of a geometrical progression applies.

In spite of all the obstacles that Teilhard perhaps wisely puts in our way, it is possible to discern a train of thought in *The Phenomenon of Man*. It is founded upon the belief that the fundamental process or motion in the entire universe is *evolution*, and evolution is 'a general condition to which all theories, all hypotheses, all systems must bow . . . a light illuminating all facts, a curve that all lines must follow'. This being so, it follows that 'nothing could ever burst forth as final across the different thresholds successively traversed by evolution . . . which has not already existed in an obscure and primordial way' (again my romans). Nothing is wholly new: there is always some primordium or rudiment or archetype of whatever exists or has existed. Love, for example – 'that is to say, the affinity of being with being' – is to be found in some form throughout the organic world, and even at a 'prodigiously rudimentary level', for if there were no such affinity between atoms when they unite into molecules it would be 'physically impossible for love to appear higher up, with us, in "hominized" form'. But above all, consciousness is not new, for this would contradict the evolutionary axiom; on the contrary, we are 'logically forced to assume the existence in rudimentary form . . . of some sort of psyche in every corpuscle', even in molecules; 'by the very fact of the individualization of our planet, a certain mass of elementary consciousness was originally imprisoned in the matter of earth'.

What form does this elementary consciousness take? Scientists have not been able to spot it, for they are shallow superficial fellows, unable to see into the inwardness of things – 'up to now, has science ever troubled to look at the world other than from *without*?' Consciousness is an interiority of matter, an 'inner face that everywhere duplicates the "material" external face, which alone is commonly considered by science'. To grasp the nature of the within of things we must understand that energy is of two kinds: the 'tangential', which is energy as scientists use that word, and a radial energy (a term used interchangeably with spiritual or psychic energy), of which consciousness is treated sometimes as the equivalent, sometimes as the manifestation, and sometimes as the consequence (there is no knowing what Teilhard intends). Radial energy appears to be a measure of, or that which conduces towards, complexity or degree of arrangement; thus 'spiritual energy, by its very nature, increases in "radial" value . . . in step with the increasing chemical complexity of the elements of which it represents the inner lining'. It confers *centricity*, and 'the increase of the synthetic state of matter involves . . . an increase of consciousness'.

We are now therefore in a position to understand what evolution is (is nothing but). Evolution is 'the continual growth of . . . "psychic" or "radial" energy, in the course of duration, beneath and within the mechanical energy I called "tangential"'; evolution, then, is 'an ascent towards consciousness'. It follows that evolution must have a 'precise *orientation* and a privileged *axis*' at the topmost pole of which lies Man, born 'a direct lineal descendant from a total effort of life'.

Let us fill in the intermediate stages. Teilhard, with a penetrating insight that Sir Julian Huxley singles out for special praise, discerns that consciousness in the everyday sense is somehow associated with the possession of nervous systems and brains ('we have every reason to think that in animals too a certain inwardness exists, approximately proportional to the development of their brains'). The direction of evolution must therefore be towards cerebralisation, that is, towards becoming brainier. 'Among the infinite modalities in which the complication of life is dispersed,' he tells us, 'the differentiation of nervous tissue stands out . . . as a significant transformation. *It provides a direction*; and by its consequences *it proves that evolution has a direction*.' All else is equivocal and insignificant; in the process of becoming brainier we find 'the very essence of complexity, of essential metamorphosis'. And if we study the evolution of living things, organic evolution, we shall find that in every one of its lines, except only in those in which it does not occur, evolution is an evolution towards increasing complexity of the nervous system and cerebralisation. Plants don't count, to be sure (because 'in the vegetable kingdom we are unable to follow along a nervous system the evolution of a psychism obviously remaining diffuse'), and the contemplation of insects provokes a certain shuffling of the feet;[1] but primates are 'a phylum of *pure and direct cerebralization*' and among them 'evolution went straight to work on the brain, neglecting everything else'. Here is Teilhard's description of noogenesis, the birth of higher consciousness among the primates, and of the noosphere in which that higher consciousness is deployed:

By the end of the Tertiary era, the psychical temperature in the cellular world had been rising for more than 500 million years . . . When the anthropoid, so to speak, had been brought 'mentally' to boiling-point some further calories were added . . . No more was needed for the whole inner equilibrium to be upset . . . By a tiny 'tangential' increase, the 'radial' was turned back on itself and so to speak took an infinite leap forward. Outwardly, almost nothing in the organs had changed. But in depth, a great revolution had taken place: consciousness was now leaping and boiling in a space of super-sensory relationships and representations . . .

The analogy, it should be explained, is with the vaporisation of water when it is brought to boiling-point, and the image of hot vapour remains when all else is forgotten.

I do not propose to criticise the fatuous argument I have just outlined; here, to expound is to expose. What Teilhard seems to be trying to say is that evolution is often (he says always) accompanied by an increase of orderliness or internal coherence or degree of integration. In what sense is the fertilised egg that develops into an adult human being 'higher' than, say, a bacterial cell? In the sense that it contains richer and more complicated genetical instructions for the execution of those processes that together constitute development. Thus Teilhard's radial, spiritual or psychic energy may be equated to 'information' or 'information content' in the sense that has been made

[1] p. 153.

reasonably precise by modern communications engineers. To equate it to consciousness, or to regard degree of consciousness as a measure of information content, is one of the silly little metaphysical conceits I mentioned in an earlier paragraph. Teilhard's belief, enthusiastically shared by Sir Julian Huxley, that evolution flouts or foils the second law of thermodynamics is based on a confusion of thought; and the idea that evolution has a main track or privileged axis is unsupported by scientific evidence.

Teilhard is widely believed to have rejected the modern Mendelian-Darwinian theory of evolution or to have demonstrated its inadequacy. Certainly he imports a ghost, the entelechy or *élan vital* of an earlier terminology, into the Mendelian machine; but he seems to accept the idea that evolution is probationary and exploratory and mediated through a selective process, a 'groping', a 'billionfold trial and error'; 'far be it from me', he declares, 'to deny its importance'. Unhappily Teilhard has no grasp of the real weakness of modern evolutionary theory, namely its lack of a complete theory of variation, of the origin of *candidature* for evolution. It is not enough to say that 'mutation' is ultimately the source of all genetical diversity, for that is merely to give the phenomenon a name: mutation is so defined. What we want, and are very slowly beginning to get, is a comprehensive theory of the forms in which new genetical information comes into being. It may, as I have hinted elsewhere, turn out to be of the nature of nucleic acids and the chromosomal apparatus that they tend spontaneously to proffer genetical variants – genetical solutions of the problem of remaining alive – which are more complex and more elaborate than the immediate occasion calls for; but to construe this 'complexification' as a manifestation of consciousness is a wilful abuse of words.

Teilhard's metaphysical argument begins where the scientific argument leaves off, and the gist of it is extremely simple. Inasmuch as evolution is the fundamental motion of the entire universe, an ascent along a privileged and necessary pathway towards consciousness, so it follows that our present consciousness must 'culminate forwards in some sort of supreme consciousness'. In expounding this thesis, Teilhard becomes more and more confused and excited and finally almost hysterical. The Supreme Consciousness, which apparently assimilates to itself all our personal consciousnesses, is, or is embodied in, 'Omega'

or the Omega-point; in Omega 'the movement of synthesis culminates'. Now Omega is 'already in existence and operative at the
very core of the thinking mass', so if we have our wits about us we
should at this moment be able to detect Omega as 'some excess of
personal, extra-human energy', the more detailed contemplation of
which will disclose the Great Presence. Although already in existence,
Omega is added to progressively: 'All round us, one by one, like a
continual exhalation, "souls" break away, carrying upwards their
incommunicable load of consciousness', and so we end up with 'a
harmonized collectivity of consciousnesses equivalent to a sort of
super-consciousness'.

Teilhard devotes some little thought to the apparently insuperable
problem of how to reconcile the persistence of individual consciousnesses with their assimilation to Omega. But the problem yields to the
application of 'remorseless logic'. The individual particles of consciousness do not join up any old how, but only centre to centre, thanks to the
mediation of Love; Omega, then, 'in its ultimate principle, can only be
a distinct Centre radiating at the core of a system of centres', and the
final state of the world is one in which 'unity coincides with a paroxysm
of harmonized complexity'. And so our hero escapes from his dire
predicament: with one bound Jack was free.

Although elsewhere Teilhard has dared to write an equation so
explicit as 'Evolution=Rise of Consciousness' he does not go so far
as to write 'Omega=God'; but in the course of some obscure pious
rant he does tell us that God, like Omega, is a 'Centre of centres',
and in one place he refers to 'God-Omega'.

How have people come to be taken in by *The Phenomenon of Man*? We
must not underestimate the size of the market for works of this kind,
for philosophy-fiction. Just as compulsory primary education created a
market catered for by cheap dailies and weeklies, so the spread of
secondary and latterly of tertiary education has created a large population of people, often with well-developed literary and scholarly tastes,
who have been educated far beyond their capacity to undertake
analytical thought. It is through their eyes that we must attempt to see
the attractions of Teilhard, which I shall jot down in the order in which
they come to mind.

1   *The Phenomenon of Man* is anti-scientific in temper (scientists are shown up as shallow folk skating about on the surface of things), and, as if that were not recommendation enough, it was written by a scientist, a fact which seems to give it particular authority and weight. Laymen firmly believe that scientists are one species of person. They are not to know that the different branches of science require very different aptitudes and degrees of skill for their prosecution. Teilhard practised an intellectually unexacting kind of science in which he achieved a moderate proficiency. He has no grasp of what makes a logical argument or of what makes for proof. He does not even preserve the common decencies of scientific writing, though his book is professedly a scientific treatise.

2   It is written in an all but totally unintelligible style, and this is construed as prima-facie evidence of profundity. (At present this applies only to works of French authorship; in later Victorian and Edwardian times the same deference was thought due to Germans, with equally little reason.) It is because Teilhard has such wonderful *deep* thoughts that he's so difficult to follow – really it's beyond my poor brain but doesn't that just *show* how profound and important it must be?

3   It declares that Man is in a sorry state, the victim of a 'fundamental anguish of being', a 'malady of space–time', a sickness of 'cosmic gravity'. The Predicament of Man is all the rage now that people have sufficient leisure and are sufficiently well fed to contemplate it, and many a tidy literary reputation has been built upon exploiting it; anybody nowadays who dared to suggest that the plight of man might not be wholly desperate would get a sharp rap over the knuckles in any literary weekly. Teilhard not only diagnoses in everyone the fashionable disease but propounds a remedy for it – yet a remedy so obscure and so remote from the possibility of application that it is not likely to deprive any practitioner of a living.

4   *The Phenomenon of Man* was introduced to the English-speaking world by Sir Julian Huxley, who, like myself, finds Teilhard somewhat difficult to follow ('If I understood him aright'; 'here his thought is not fully clear to me'; etc.).[1] Unlike myself, Sir Julian finds Teilhard in possession of a 'rigorous sense of values', one who 'always endeav-

[1] p. 16 and again p. 18; p. 19.

oured to think concretely'; he was speculative, to be sure, but his speculation was 'always disciplined by logic'. But then it does not seem to me that Huxley expounds Teilhard's argument; his Introduction does little more than to call attention to parallels between Teilhard's thinking and his own. Chief among these is the cosmic significance attached to a suitably generalised conception of evolution – a conception so diluted or attenuated in the course of being generalised as to cover all events or phenomena that are not immobile in time.[1] In particular, Huxley applauds the, in my opinion, mistaken belief that the so-called 'psychosocial evolution' of mankind and the genetical evolution of living organisms generally are two episodes of a continuous integral process (though separated by a 'critical point', whatever that may mean). Yet for all this Huxley finds it impossible to follow Teilhard 'all the way in his gallant attempt to reconcile the supernatural elements in Christianity with the facts and implications of evolution'. But, bless my soul, this reconciliation is just what Teilhard's book is *about*!

I have read and studied *The Phenomenon of Man* with real distress, even with despair. Instead of wringing our hands over the Human Predicament, we should attend to those parts of it which are wholly remediable, above all to the gullibility which makes it possible for people to be taken in by such a bag of tricks as this. If it were an innocent, passive gullibility it would be excusable; but all too clearly, alas, it is an active willingness to be deceived.

[1] pp. 12, 13.

THE ACT OF CREATION

THE author of *Darkness at Noon* must be listened to attentively, no matter what he may choose to write upon. Arthur Koestler is a very clever, knowledgeable, and inventive man, and *The Act of Creation*[1] is very clever too, and full of information, and quite wonderfully inventive in the use of words. Many of the points it makes are not likely to be challenged. That wit and creative thought have much in common; that great syntheses may be come upon by logically unmapped pathways; that putting two and two together is an important element in discovery and also, in a certain sense, in making jokes: it has all been said before, of course, and in fewer than 750 pages, though never with such vitality; and anyhow much of it will bear repeating. But as a serious and original work of learning I am sorry to say that, in my opinion, *The Act of Creation* simply won't do. This is not because of its amateurishness, which is more often than not endearing, nor even because of its blunders – they don't affect Koestler's arguments very much one way or another, even when they reveal a deep-seated misunderstanding of, for example, 'Neo-Darwinism', or find expression in fatuous epigrams like 'All automatic functions of the body are patterned by rhythmic pulsations.' I shall try to explain later what I think wrong with Koestler's technical arguments, but let us first of all examine *The Act of Creation* at the level of philosophical *belles lettres*.

As to style, Koestler overdoes it. On one half-page catharsis is described as an 'earthing' of the emotions, the satisfaction of seeing a joke is said to supply 'added voltage to the original charge detonated in laughter', and a smutty story is put at 'the infra-red end of the emotive spectrum'. We aren't quite sure when he intends to be taken literally: for example, what about 'A concept has as many dimensions in semantic space as there are matrices of which it is a member'? This could have been intended to express an exact idea, for 'space' and 'dimension' have generalised technical meanings; but the feeble passage

[1] (London, 1964).

that follows, illustrated by the word 'Madrid', tells us only that the word itself and the city it stands for conjure up all kinds of different associations in his mind.

When I started Koestler's book I hoped he was going to do something for which his knowledge and sympathies and writer's insight should give him unequalled qualifications: that he would give us a first ethology of scientific activity, and so help to make the scientist intelligible to others and to himself. Unhappily, there are passages in *The Act of Creation* which convince me that he has no real grasp of how scientists go about their work. Consider the depths of misunderstanding revealed by Koestler's aloof and snobbish remarks about the 'unseemly haste' with which some scientists publish their discoveries. Koestler's historical hobnobbings with men of genius seem to have made him forget the fact that, in science, what X misses today Y will surely hit upon tomorrow (or maybe the day after tomorrow). Much of a scientist's pride and sense of accomplishment turns therefore upon being the *first* to do something – upon being the man who did actually speed up or redirect the flow of thought and the growth of understanding. There is no spiritual copyright in scientific discoveries, unless they should happen to be quite mistaken. Only in making a blunder does a scientist do something which, conceivably, no one else might ever do again. Artists are not troubled by matters of priority, but Wagner would certainly not have spent twenty years on *The Ring* if he had thought it at all possible for someone else to nip in ahead of him with *Götterdämmerung*.

Like other amateurs, Koestler finds it difficult to understand why scientists seem so often to shirk the study of really fundamental or challenging problems. With Robert Graves he regrets the absence of 'intense research' upon variations in the – ah – 'emotive potentials of the sense modalities'. He wonders why 'the genetics of behaviour' should still be 'uncharted territory' and asks whether this may not be because the framework of neo-Darwinism is too rickety to support an enquiry. The real reason is so much simpler: the problem is very, very difficult. Goodness knows how it is to be got at. It may be outflanked or it may yield to attrition, but probably not to direct assault. No scientist is admired for failing in the attempt to solve problems that lie beyond his competence. The most he can hope for is the kindly contempt earned by the Utopian politician. If politics is the art of the possible, research is surely the art of the soluble. Both are immensely practical-minded affairs.

Although much of Koestler's book has to do with explanation, he seems to pay little attention to the narrowly scientific usages of the concept. Some of the 'explanations' he quotes with approval[1] are simply analgesic pills which dull the aches of incomprehension without going to their causes. The kind of explanation the scientist spends most of his time thinking up and testing – the hypothesis which enfolds the matters to be explained among its logical consequences – gets little attention. Instead we are told that there are all kinds of explanations and many degrees of understanding, starting with the '*unconscious* understanding mediated by the dream'.

Dreams bring out the worst in Koestler. Dreaming is a 'sliding back towards the pulsating darkness, one and undivided, of which we were part before our separate egos were formed'. No wonder, then, that the understanding it conveys is of the unconscious kind. 'There is no need to emphasize, in this century of Freud and Jung, that the logic of the dream . . . derives from the magic type of causation found in primitive societies and the fantasies of childhood.' But those who enjoy slopping around in the amniotic fluid should pause for a moment to entertain (perhaps only unconsciously in the first instance) the idea that the content of dreams may be totally devoid of 'meaning'. There should be no need to emphasise, in this century of radio sets and electronic devices, that many dreams may be assemblages of thought-elements that convey no information whatsoever: that they may just be *noise*.[2]

Koestler's theory of the creative act is set out in Book One as a special theory comprehended within a general theory that occupies Book Two. In Book One he defines two special notions, 'matrix' and 'code'. A code is a system of rules of process or performance and a matrix is 'any ability, habit, or skill, any pattern of ordered behaviour' governed by a code. In particular, a matrix of thought is, or can for

[1] For example, his account (p. 452) of C. M. Child's explanations of 'physiological isolation'.

[2] 'Noise' as the word is used by communication theorists: I borrowed this thought from an impromptu of Professor A. L. Hodgkin's.

variety's sake be described as, a 'frame of reference', an 'associative context', a 'type of logic', or a 'universe of discourse'. Behind every act of creation lies a binary association (bisociation) of matrices: in that which provokes laughter they collide, in a new intellectual synthesis they fuse, and in an aesthetic experience they confront each other or are juxtaposed. These three degrees of experience form a continuum, and may indeed grow out of the bisociation of the very same matrices. Thus the exuberant, explosive, tension-relieving delight (*Eureka!*) and the long after-glow ('the oceanic feeling') excited by an intellectual synthesis have their counterparts in laughter and in the sense of satisfaction at seeing a joke.

Koestler is far too intelligent a man not to realise that his account of creative activity is full of difficulties, but though he mentions the contexts in which some of them arise, he does not direct attention to them explicitly or make any attempt to work them out. Among them, and in no special order, are: (*a*) Just how does an explanation which later proves false (as most do: and none, he admits, is proved true) give rise to just the same feelings of joy and exaltation as one which later stands up to challenge? What went wrong: didn't the matrices fuse, or were they the wrong kind of matrix, or what? (*b*) The source of most joy in science lies not so much in devising an explanation as in getting the results of an experiment which upholds it. (*c*) Some awkward problems are raised by the fact that the chap who *sees* a joke splits his sides laughing as well, maybe, as the chap who makes it; but the chap who 'sees' or is apprised of an intellectual synthesis does not share in the tension-relieving, explosive joy of discovery. Likewise the joy of artistic creation 'travels' in some sense in which the joy of intellectual synthesis does not, and this difference between them seems to me to outweigh their similarities. (*d*) It follows from Koestler's scheme that an intellectual synthesis, upon being proved false, should at once become a huge joke, especially to the person who devised it, for it must have rested upon the kind of bisociative act that underlies the comical. However, we are not amused. (*e*) The sense of comfort an explanation may give rise to has nothing to do with bringing about or even witnessing a 'fusion of matrices': laymen get it not so much from knowing an explanation as from knowing that an explanation is known.

I should mention, because Koestler does not, that the so-called 'hypothetico-deductive' interpretation of the scientific process copes perfectly with difficulties (a) and (b), and that in it (c), (d) and (e) do not arise. Devising a hypothesis is a 'creative act' in the sense that it is the invention of a possible world, or a possible fragment of the world; experiments are then done to find out whether or not that imagined world is, to a good enough approximation, the real one. As Koestler conceives it, the act of creation is not, in the usual sense, creative at all; as he says, it merely 'uncovers, selects, reshuffles, combines, synthesises already existing facts, ideas, faculties, skills'.

Koestler's psychological thought, though not confessedly 'introspective', is in the style of the nineteenth century – a point delicately made, as I read it, by Sir Cyril Burt in his foreword. Koestler nags away at behaviourism, which he describes as 'the dominant school in contemporary psychology', though later he says of J. B. Watson's textbook that 'few students today remember its contents, or even its basic postulates'. For all its crudities behaviourism, conceived as a methodology rather than as a psychological system, taught psychology with brutal emphasis that 'The dog is whining' and 'The dog is sad' are statements of altogether different empirical standing, and heaven help psychology if it ever again overlooks the distinction.

I was dreading the moment in my reading of Koestler when I was to be told that sexual reproduction, 'the bisociation of two genetic codes', was 'the basic model of the creative act'. Koestler's Book Two contains the general theory which comprehends the special theory I have just outlined. It is full of rather old-fashioned biology (but what fun to read again of axial gradients!), and the person who has to be told on page 57 that the sympathetic nervous system has nothing to do with the emotion of sympathy will not make much of it. The argument runs thus. The system of nature is a hierarchical system of elements, sub-elements, sub-sub-elements etc., each enjoying a certain wholeness and autonomy, but each also subordinate to the element above it in the hierarchy. The structure of army command is one example of such a hierarchy (a company has some autonomy but is subordinate to the battalion) and in the living world the hierarchy of organism, organ, cell, cell part etc. is another. At each level we find a certain wholeness and a certain partness.

Koestler declares that at each level of the hierarchy 'homologous' principles operate, with the consequence that any phenomenon at one level must have its homologue or formal counterpart at each other level. In particular, we shall find 'mental equivalents' of what goes on at lower levels in the hierarchy, and conversely, since we can go either up or down the ladder of correspondences, physical equivalents of what goes on in the mind. The 'creative stress' of the artist or scientist corresponds to the 'general alarm reaction' of the injured animal, and dreaming is the mental equivalent of the regenerative processes that make good wear and tear. Embryonic development has a certain self-assertive quality, and so have 'perceptual matrices'. A not yet verbalised analogy corresponds to an organ rudiment in the early embryo, and rhythm and rhyme, assonance and pun, are 'vestigial echoes' of the 'primitive pulsations of living matter'.

No metaphors these: 'they have solid roots in the earth'; but to my ears they sound silly, and I believe them to be as silly as they sound. Disregarding the rights and wrongs of building hierarchies out of non-homogeneous elements (though in fact it won't do to mix up perceptual matrices with adrenal glands, embryos, jokes and rhymes), the correspondences Koestler makes so much of are of the purely formal and abstract kind that can be expressed without any regard to their empirical content. Even if the networks of relationships holding at each level of the hierarchy were isomorphic, there would be no necessary affinity between the things so related. The correspondences which Koestler urges us to believe in are harmless enough, but arguments of this kind can be mischievous (for example, a case for totalitarian government can be built upon an unsound analogy between organism and State).

Koestler makes a good point when he says that during the past hundred years or so scientists have felt themselves under a professional obligation to write in a dry, cold, pulseless way; to be, in short, boring. (It is part of the heritage of inductivism.) *The Act of Creation* is so full of vitality that it creates around it an aura of good-to-be-alive, and though Koestler regards himself as the author of a new and important general psychological theory, I am delighted that, in writing a 'popular' work, he has in a sense appealed to the general public over the heads of the profession. But certain rules of scientific manners must be observed no matter what form the account of a scientific theory may take. One must mention (if only to dismiss with contempt) other, alternative explanations of the matters one is dealing with; and one must discuss (if only to prove them groundless) some of the objections that are likely to be raised against one's theories by the ignorant or ill-disposed. Koestler seems to have no adequate grasp of the importance of *criticism* in science – above all of self-criticism, for most of a scientist's wounds are self-inflicted. Nor can I remember in his book any passage suggesting observations or experiments which might qualify or refine his ideas. He quotes with approval one 'laconic pronouncement' of Dirac's which must have made sense in context but which otherwise sounds just naughty: 'It is more important to have beauty in one's equations than to have them fit experiment.' The high inspirational origin of a theory is no guarantee of its trustworthiness, and Koestler should avoid giving the impression that he thinks it is. No belief could bring science more quickly to ruin.

Mr Koestler replied thus:

SIR – Allow me to answer some of the points raised in Professor Medawar's review of my recent book.

  1 Medawar writes:

There are passages in *The Act of Creation* which convince me that he has no real grasp of how scientists go about their work. Consider the depths of misunder-standing revealed by Koestler's aloof and snobbish remarks about the 'unseemly haste' with which some scientists publish their discoveries. Koestler's historical hobnobbings with men of genius seem to have made him forget the fact that, in science, what X misses today Y will surely hit upon tomorrow . . .

The snobbish remarks to which this passage refers read as follows:

In 1922, Ogburn and Thomas published some 150 examples of discoveries and inventions which were made independently by several persons; and, more recently, Merton came to the seemingly paradoxical conclusion that 'the pattern of independent multiple discoveries in science is . . . the dominant pattern rather than a subsidiary one'. He quotes as an example Lord Kelvin, whose published papers contain 'at least thirty-two discoveries of his own which he subsequently found had also been made by others . . .'. The endless priority disputes which have poisoned the supposedly serene atmosphere of scientific research throughout the ages, and the unseemly haste of many scientists to establish priority by rushing into print – or, at least, depositing manuscripts in sealed envelopes with some learned society – point in the same direction. Some – among them Galileo and Hooke – even went to the length of publishing half-completed discoveries in the form of anagrams, to ensure priority without letting rivals in on the idea.

2 Medawar seems to object to my quite unoriginal contention that unconscious processes in the dream and in the hypnagogic state between dreaming and awakening often play a decisive part in scientific discovery. At least this seems to be the meaning behind the heavy veils of irony in the passage:

Dreams bring out the worst in Koestler . . . those who enjoy slopping around in the amniotic fluid should pause for a moment to entertain (perhaps only unconsciously in the first instance) the idea that the content of dreams may be totally devoid of 'meaning' . . . that many dreams may be assemblages of thought-elements that convey no information whatsoever.

No doubt most dreams are self-addressed messages whose information-content is purely private and 'meaningless' to others. But equally undeniable is the fact – which Medawar chooses to pass in silence – that dreams, hypnagogic images and other forms of unconscious intuitions proved decisive in the discoveries of dozens of scientists and mathematicians whose testimonies I quoted – among them Ampère, Gauss, Kekulé, Leibnitz, Poincaré, Fechner, Otto Loewi, Planck, Einstein, to mention only a few.

3 Medawar raises five objections, numbered (a) to (e), against the theory I proposed. To save space, let me refer to the passages in the book in which the answers to these objections can be found. (a) Bk One, Chap. IX, p. 212 et seq. (b) The 'joy' in 'devising an explanation' and the satisfaction derived from its empirical confirmation enter at different stages and must not be confused. (c) Bk One, Chap. IV, p. 87 et seq. and Chap. XI, p. 255 et seq. (d) 'It follows from Koestler's scheme, etc.'. The answer is, it does not. (e) See Bk One, Chap. XVII, pp. 325-31.

4 Medawar accuses me of quoting out of context 'one "laconic pronouncement" of Dirac's'. The single sentence which Medawar requotes is on p. 329. The full context, which Medawar overlooked, is to be found on pp. 245-6. If he had no time to read through the book, he should at least have looked at the index.

5 Medawar accuses me of contradicting myself: 'Koestler nags away at behaviourism, which he describes as "the dominant school in contemporary psychology", though later he says of J. B. Watson's textbook that "few students today remember its contents or even its basic postulates".' This I said on p. 558. On p. 559 I continued:

Although the cruder absurdities of Watsonian behaviourism are forgotten, it had laid the foundations on which the later, more refined behaviouristic systems [of Guthrie, Hull, and Skinner] were built; the dominant trend in American and Russian psychology in the generation that followed had a distinctly Pavlov–Watsonian flavour. The methods became more sophisticated, but the philosophy behind them remained the same.

The rest consists of ironic innuendo and *ex cathedra* pronouncements. 'Certain rules of scientific manners must be observed,' Professor Medawar informs us. I wish he had lived up to his precept.

I in turn answered:

I should like to take Mr Koestler's points one by one.

1 *Priority.* A scientist's sense of concern about matters of priority may not be creditable, but only prigs deny its existence, and the fact that it does exist points towards something distinctive in the act of creation as it occurs in science. It is not good enough to brush it aside with clichés ('unseemly haste', 'rushing into print') or to pour scorn on its extremer manifestations. I think my own interpretation was the right one – priority in science gives moral possession – but Koestler seems not to realise that there is anything to interpret. As to simultaneous discovery: it was the slight air of wonderment about it in the very passage Koestler now quotes which made me ask if he realised the consequences of the relationship, peculiar to science, between X and Y. Simultaneous discovery is utterly commonplace, and it was only the rarity of scientists, not the inherent improbability of the phenomenon, that made it remarkable in the past. Scientists on the same road may be expected to arrive at the same destination, often not far apart. Romantics like Koestler don't like to admit this, because it seems to them to derogate from the authority of genius. Thus of Newton and Leibniz, equal first with the differential calculus, Koestler says 'the greatness of this accomplishment is hardly diminished by the fact that two

among millions, instead of one among millions, had the exceptional genius to do it'. But millions weren't *trying* for the calculus. If they had been hundreds would have got it. Very simple-minded people think that if Newton had died prematurely we should still be at our wits' end to account for the fall of apples. Is there not just a trace of this *naïveté* in Koestler?

2 *Dreams.* Koestler quite misses the point, and what he says is a good example of how stubbornly the mind may deny entry to the unfamiliar (*The Act of Creation*, p. 216). I did not suggest that dreams conveyed private messages whose import was known only to the dreamer. My proposal, as unoriginal as Koestler's, was that dreams are not messages at all. It is naughty of Koestler to lump together 'dreams, hypnagogic images and other forms of unconscious intuitions', as if my misgivings about dreams extended to all other unscripted activities of the mind. They don't: if we are to brandish texts at each other, I will cite an article in the *Times Literary Supplement*[1] in which I speak up for inspiration and against the idea that discovery can be logically mechanised.

3 *Objections (a), (c) and (e).* Koestler would not have drawn special attention to these passages in his book unless he really believed them to hold the answers to my, as I think, damaging objections to his theory. Now, on rereading him, I feel convinced that he simply doesn't understand the *kind* of intellectual performance that is expected of someone who propounds or defends a scientific or philosophic theory. But now we have both had our say and it can all go out to arbitration. (*b*) We agree, then: but why here no citations of the passages in his book in which he says so? (*d*) I'm so sorry, but I still think it follows from Koestler's scheme that an intellectual synthesis, upon being proved false, should appear funny. (If I were parodying Koestler I should describe it as a joke played by Nature of which we were very slow to see the point.) What's more, the converse also follows, that a great intellectual synthesis wrongly believed false will be thought hugely comical from the outset. Koestler seems to think so too:

Until the seventeenth century the Copernican hypothesis of the earth's motion was considered as obviously incompatible with common-sense experience; it

---

[1] 'Hypothesis and Imagination', pp. 115–35 above.

was accordingly treated as a huge joke . . . The history of science abounds with examples of discoveries greeted with howls of laughter because they seemed to be a marriage of incompatibles . . .[1]

But as I say, in real life the refutation of a hypothesis can be deeply upsetting. If there are howls they are not of laughter.

4 Dirac's allegedly 'laconic pronouncement' ('It is more important to have beauty in one's equations than to have them fit experiment') is in reality a very diffident expression of opinion whose context I was wrong to overlook. But it was rash of Koestler to draw my attention to it, for what Dirac goes on to say is: 'It seems that if one is working from the point of view of getting beauty in one's equation, *and if one has really a sound insight*, one is on a sure line of progress.' The italics are mine, but Koestler is welcome to them.

5 *Behaviourism*. No, I didn't say Koestler contradicted himself, though I do find his love–hate relationship with modern experimental psychology extremely tiresome. Nor do I think he has quite got my point, which was that even if behaviourism were dead as a system it is still very much alive as a methodology.

Koestler must have had some doubts about the wisdom and taste of his final sentences, and I suppose it will only make matters worse if I say I forgive him. I will not, however, forgive him for hinting that I didn't read his book, nor for the fact that I had to spend hours and hours and hours in doing so.

POSTSCRIPT

For completeness' sake it should be added that Mr Iain Hamilton in *Koestler: A Biography*[2] published a loyal but in my judgement philosophically inexpert attempt to defend Koestler against my criticisms of *The Act of Creation*. Reconsideration of the whole controversy makes me now more than ever regret that Koestler did not concentrate his attention and his energy upon imaginative writing of the kind in which he is so superbly proficient.

[1] pp. 94–5.   [2] (London, 1982).

J. B. S.

THE lives of scientists, considered as Lives, almost always make dull reading. For one thing, the careers of the famous and the merely ordinary fall into much the same pattern, give or take an honorary degree or two, or (in European countries) an honorific order. It could hardly be otherwise. Academics can only seldom lead lives that are spacious or exciting in a worldly sense. They need laboratories or libraries and the company of other academics. Their work is in no way made deeper or more cogent by privation, distress or worldly buffetings. Their private lives may be unhappy, strangely mixed up or comic, but not in ways that tell us anything special about the nature or direction of their work. Academics lie outside the devastation area of the literary convention according to which the lives of artists and men of letters are intrinsically interesting, a source of cultural insight in themselves. If a scientist were to cut his ear off, no one would take it as evidence of a heightened sensibility; if a historian were to fail (as Ruskin did) to consummate his marriage, we should not suppose that our understanding of historical scholarship had somehow been enriched.

The lives of writers, however, are thought to give out a low rumble of cultural portents. One day in the summer of 1968 *The Times* of London devoted a whole column on its front page to new discoveries about the circumstances under which Joseph Conrad came to be discharged by the captain of the *Riversdale*, in which he was serving as first mate – discoveries described as 'extremely important' for the understanding of Conrad and his art, because traces of the incident are to be discerned in Conrad's fiction. (But what are we to make of a scale of values in which such a discovery ranks as extremely important? Would it not have been of epoch-making significance – nay, downright interesting – if Conrad had *not* used his experience of shipboard life in writing his stories about the sea?)

Yet J. B. S. Haldane's life, as Ronald Clark recounts it,[1] is fascina-

[1] *J.B.S.: The Life and Work of J. B. S. Haldane* (London, 1968).

ting from end to end. Unless one is in the know already, there is no foretelling at one moment what comes next. Haldane had a flying start in life. His father was a famous physiologist; his uncle translated Schopenhauer and became Lord Chancellor and Minister of War; his aunt was a distinguished social reformer; and his sister Naomi Mitchison (the dedicatee, incidentally, of *The Double Helix*) is a well-known writer and, among other things, an honorary member of the Bakgatla tribe. Even the house he was brought up in has been transformed into an Oxford College.

The Dragon School, Eton and Oxford gave Haldane about the best education a man of his generation could have. At Oxford he was an authentic 'Double First', for having taken first-class honours in Mathematical Moderations, he switched to philosophy and ancient history and took a first-class degree in Greats, the most prestigious thing an Oxford undergraduate could do. Haldane could have made a success of any one of half a dozen careers – as mathematician, classical scholar, philosopher, scientist, journalist or imaginative writer. In unequal proportions he was in fact all of these. On his life's showing he could not have been a politician, administrator (heavens, no!), jurist or, I think, a critic of any kind. In the outcome he became one of the three or four most influential biologists of his generation.

In some respects – quickness of grasp, and the power to connect things in his mind in completely unexpected ways – he was the cleverest man I ever knew. He had something novel and theoretically illuminating to say on every scientific subject he chose to give his mind to: on the kinetics of enzyme action, on infectious disease as a factor in evolution, on the relationship between antigens and genes, and on the impairment of reasoning by prolonged exposure to high concentrations of carbon dioxide. Haldane was the first to describe the genetic phenomenon of linkage in animals generally, and the first to estimate the mutation rate in man. His greatest work began in the 1920s, when independently of Sewall Wright and R. A. Fisher he undertook to refound Darwinism upon the concepts of Mendelian genetics. It should have caused a great awakening of Darwinian theory, and in due course it did so; but at the time it did no more than make Darwinism stir in its dogmatic slumbers, and even today, on the Continent, what passes for Darwinism is essentially the Darwinism of fifty years ago. (The same goes for the neo-Darwinism denounced by modern nature-philosophers, who are handicapped by the fact that the mathematical theory of natural selection is too difficult for them to understand.)

The gist of the newer evolutionism is as follows. In principle, every living organism can be allotted a formula representing its genetic constitution, its make-up in terms of 'genes'. In a natural population of some one species, each gene represented in the population will have a certain frequency of occurrence, because it will often occur in some members but not in others, and the population considered as a whole will therefore itself have a certain genetic make-up, definable in terms of the frequency of genes. In 1908 the mathematician G. H. Hardy announced the following fundamental theorem: that although, through the workings of Mendelian heredity, a virtually infinite variety of new combinations and reassortments of genes will turn up in members of the population, yet if mating is random, and all organisms have an equal chance of leaving offspring, the frequency of each gene in the population will necessarily remain constant from one generation to the next. We must therefore look to some impressed 'force' acting over the population as a whole if the gene frequency is to change in some systematic (that is, other than merely random) way – if, in short, the population is to evolve. The most important of these forces is natural selection, a compendious name for all the agencies that cause one section of the population to make a disproportionately large contribution to the ancestry of future generations. The rate of change of gene frequency is therefore a measure of the magnitude of the force of natural selection.

In the light of this new conception, Haldane, Fisher and Wright were able for the first time to describe the phenomenon of evolution in a genetic language, and to reveal the delicacy and subtlety of a population's response to selective forces. In this scheme of thinking, 'mutation' occupies a certain special place. Mutation adds to genetic diversity, and therefore enlarges the candidature for evolutionary change. This is quite different from saying (as older Darwinians and modern nature-philosophers say) that the mutant organism is itself the candidate for evolution, the hopeful variant that is selected or rejected as the case may be. The newer evolution theory represents a revolution of thought of the same general kind and the same stature as that which led to the development of statistical mechanics in the latter half of the nineteenth century.

This then is 'classical' work, assimilated into all the standard text-books. If he had done nothing else, Haldane would still be classified as one of the grand masters of modern evolution theory. Yet he was not a profoundly original thinker. His genius was to enrich the soil, not to bring new land into cultivation. He was not himself the author of any great new biological conception, nor did his ideas arouse the mis-givings and resentment so often stirred up by what is revolutionary or profoundly original. On the contrary, everything he said was at once recognised as fruitful and illuminating, something one would have been proud and delighted to have thought of oneself, even if later research should prove it to be mistaken.

Haldane had a reputation for being a bad experimenter, anyhow in the narrower, manipulative sense. I never saw him in the traditional laboratory uniform of white or once-white coat, and he gave the im-pression of being clumsy with his hands. He could design experiments, of course, and guide others in their execution, but he was not by nature an experimentalist; he did not translate scientific problems into a lan-guage of conjecture and refutation. His great strength was to see con-nections, to put two and two together, to work out the deeper or re-moter consequences of taking certain theoretical views. If he had been a physicist he would have been a theoretical physicist (and in a small way he actually was), but experimentalists liked talking to him ('Let's try it out on the Prof.'), and only the obtuse could have failed to derive benefit from what he said.

Between his First in Greats in 1914 and his return to Oxford in 1919 to study physiology, Haldane was away at the wars. I knew Haldane only during the latter half of his life, and had not realised until I read Clark's biography how thoroughly Haldane enjoyed everything that went with war. The First World War seems to have been the happiest period of Haldane's life; we have his word for it that he disliked Eton, and he was not yet victim of the many vexations, real or imagined, that took the edge off his enjoyment of professional life. Haldane described life in the front line as 'truly enviable'; he enjoyed the com-radeship of war and even (if what he says is anything to go by) the experience of killing people: 'I get a definitely enhanced sense of life when my life is in moderate danger.' Courage he disclaimed, but he was to all appearances fearless. His bravery, as I construe it, was the product of a superb intellectual arrogance – a complete confidence in the accuracy of his assessment of degrees of risk. Like Houdini, he judged safe the exploits that to others seemed suicidal: 'I once bicycled across a gap in full view of the Germans, having foreseen that they would be too surprised to open fire.' When the Germans started using chlorine in 1915, Haldane was taken out of the front line to help his father study its physiological effects – research of the utmost importance which showed up Haldane at his very best, and carried out at some risk (which he underestimated) to himself and his physiological colleagues.

What are we to make of Haldane as a human being? The first thing to be said in answer to such a question is that we are under no obligation to make anything of him at all. It makes no difference now. It might have made a difference if Haldane in his lifetime could have been made to realise the degree to which his work was obstructed by his own perversity. He was so ignorant of anything to do with administration that he did not even know how to call the authorities' attention to the contempt in which he held them. When he burst into terrible anger about his grievances, it was over the heads of minor functionaries and clerks. The cleaners were terrified of him, and the electricians were said to have demanded danger money for working in his room. His room was therefore never cleaned; it became a sort of showpiece, littered with fossil specimens undergoing a second interment.

Clark describes a scene which those who knew him came to regard as typical. On behalf of one of his students Haldane applied for one of the Agricultural Research Council's postgraduate awards. These awards are made provisionally, and are confirmed if the candidate gets a high enough degree. To speed things, one of the Council's junior officers rang up Haldane's secretary to find out what class of degree the candidate had in fact been given. As it happened, the class lists had not yet been published, so the reasonable answer would have been 'I'm sorry, we can't tell you yet because the results aren't out.' Instead Haldane accused the Council of blackmail and an attempt to violate the secrecy of exams: 'I refuse to give you the information, and withdraw my request for a grant. I shall pay for her out of my own pocket.'

Yet more than once he scored an important victory: over the *Sex Viri*, for example, a sort of *buffo* male-voice sextet that tried to deprive him of his readership at Cambridge on the grounds of immorality. Indeed, the scenes accompanying the divorce that freed Charlotte Burghes to become Haldane's first wife read like the libretto of a comic opera, including an adultery that was chaste in spite of all appearances to the contrary.

In America, Haldane was notorious for his communist professions: he was ideologically a communist during the latter part of his life, joining the party officially in 1942 and leaving it furtively around 1948, though he continued to write for the *Daily Worker* until 1950. Clark quotes a draft letter of resignation from the party written in 1948, but perhaps never posted. The reasons he gave for wanting to leave the party were so utterly trivial – a squabble about royalties, and various accusations of bad faith – as to make one question the seriousness and solemnity of the motives that led him to join it in the first place. In his public professions Haldane was the complete party man. Lysenko he thought 'a very fine biologist', so Clark tells us; and I know from conversation with him that he thought it quite likely that Beria, then lately disgraced, had been in the pay of the Americans, and that Slansky and Clementis, the victims of ritual hangings in Czechoslovakia, had got the punishment they deserved.

People were wont to ask how such a clever man could be so completely taken in by Communist propaganda, but Haldane was not clever in respect of any faculty that enters into political judgement. He was totally lacking in worldly sense, a sulky innocent, a wholehearted believer in *Them* – the agents of that hidden conspiracy against ordinary decent people, the authorities who withheld the grants he had never asked for and who broke the promises they had never made.

We must not take all Haldane's protestations at their face value. His declaration that he left England to live in India because of the disgrace of Suez was an effective way of expressing his contempt for the Suez adventure, but it simply wasn't true. I remember Haldane's once going back on a firm promise to chair a lecture given by a distinguished American scientist on the grounds that it would be too embarrassing for the lecturer: he had once been the victim of a sexual assault by the lecturer's wife. The accusation was utterly ridiculous, and Haldane did not in the least resent my saying so. He didn't want to be bothered with the chairmanship, and could not bring himself to say so in the usual way. But the trouble was that his extravagances became self-defeating. He became a 'character', and people began laughing in anticipation of what he would say or be up to next. It is a sort of Anglo-Saxon form of liquidation, more humane but politically not much less effective than the form of liquidation he condoned. In the Russia of Haldane's day, as Mr Clark makes clear, Haldane would have been much more offensive and with very much better reasons, but he would not have lasted anything like so long.

Physical bravery, but sometimes moral cowardice; intelligence and folly, reasonableness and obstinacy, kindness and aggressiveness, generosity and pettiness – it is like a formulary for all mankind: Haldane was a Johnsonian figure, a with-knobs-on variant of us all; but unless we are bored with life or altogether fed up with human beings, we shall not tire of reading how lofty thoughts can go with silly opinions, or how a man may strive for freedom and yet sometimes condone the work of its enemies.

LUCKY JIM

ON 30 May 1953 James Watson and Francis Crick published in *Nature* a correct interpretation of the crystalline structure of deoxyribonucleic acid, DNA. It was a great discovery, one which went far beyond merely spelling out the spatial design of a large, complicated and important molecule. It explained how that molecule could serve genetic purposes – that is to say, how DNA, within the framework of a single common structure, could exist in forms various enough to encode the messages of heredity. It explained how DNA could be stable in a crystalline sense and yet allow for mutability. Above all it explained in principle, at a molecular level, how DNA undergoes its primordial act of reproduction, the making of more DNA exactly like itself. The great thing about their discovery was its completeness, its air of finality. If Watson and Crick had been seen groping towards an answer; if they had published a partly right solution and had been obliged to follow it up with corrections and glosses; if the solution had come out piecemeal instead of in a blaze of understanding: then it would still have been a great episode in biological history, but something more in the common run of things; something splendidly well done, but not done in the grand romantic manner.

The work that ended by making biological sense of the nucleic acids began forty years ago in the shabby laboratories of the Ministry of Health in London. In 1928 Dr Fred Griffith, one of the Ministry's Medical Officers, published in the *Journal of Hygiene* a paper describing strange observations on the behaviour of pneumococci – behaviour which suggested that they could undergo something akin to a trans-mutation of bacterial species. The pneumococci exist in a variety of genetically different 'types', distinguished one from another by the chemical make-up of their outer sheaths. Griffith injected into mice a mixture of dead pneumococcal cells of one type and living cells of another type, and in due course he recovered living cells of the type that distinguished the dead cells in the original mixture. On the face of it, he had observed a genetic transformation. There was no good reason to question the results of the experiment. Griffith was a well known and highly expert bacteriologist whose whole professional life had been devoted to describing and defining the variant forms of bacteria, and his experiments (which forestalled the more obvious objections to the meaning he read into them) were straightforward and convincing. Griffith, above all an epidemiologist, did not follow up his work on pneumococcal transformation; nor did he witness its apotheosis, for in 1941 a bomb fell in Endell Street which blew up the Ministry's laboratory while he and his close colleague William Scott were working in it.

The analysis of pneumococcal transformations was carried forward by Martin Dawson and Richard Sia in Columbia University and by Lionel Alloway at the Rockefeller Institute. Between them they showed that the transformation could occur during cultivation outside the body, and that the agent responsible for the transformation could pass through a filter fine enough to hold back the bacteria themselves. These experiments were of great interest to bacteriologists because they gave a new insight into matters having to do with the ups and downs of virulence; but most biologists and geneticists were completely unaware that they were in progress. The dark ages of DNA came to an end in 1944 with the publication from the Rockefeller Institute of a paper by Oswald Avery and his young colleagues, Colin MacLeod and Maclyn McCarty, which gave very good reasons for supposing that the transforming agent was 'a highly polymerized and viscous form of sodium deoxyribonucleate'. This interpretation aroused much resentment, for many scientists unconsciously deplore the resolution of mysteries they have grown up with and have therefore come to love. It nevertheless withstood all efforts to unseat it. Geneticists marvelled at its significance, for the agent that brought about the transformation could be thought of as a naked gene. So very probably the genes were not proteins after all, and the nucleic acids themselves could no longer be thought of as a sort of skeletal material for the chromosomes.

The new conception was full of difficulties, the most serious being that (compared with the baroque profusion of different kinds of proteins) the nucleic acids seemed too simple in make-up and too little variegated to fulfil a genetic function. These doubts were set

at rest by Crick and Watson: the combinatorial variety of the four
different bases that enter into the make-up of DNA is more than
enough to specify or code for the twenty different kinds of amino
acids of which proteins are compounded; more than enough, indeed,
to convey the detailed genetic message by which one generation of
organisms specifies the inborn constitution of the next. Thanks to the
work of Crick and half a dozen others, the form of the genetic code,
the scheme of signalling, has now been clarified, and thanks to work
to which Watson has made important contributions, the mechanism
by which the genetic message is mapped into the structure of a protein
is now in outline understood.

It is simply not worth arguing with anyone so obtuse as not to
realise that this complex of discoveries is the greatest achievement
of science in the twentieth century. I say 'complex of discoveries'
because discoveries are not a single species of intellection; they are
of many different kinds, and Griffith's and Crick-and-Watson's were
as different as they could be. Griffith's was a synthetic discovery, in the
philosophic sense of that word. It did not close up a visible gap in
natural knowledge, but entered upon territory not until then known
to exist. If scientific research had stopped by magic in, say, 1920, our
picture of the world would not be *known* to be incomplete for want of
it. The elucidation of the structure of DNA was analytical in character.
Ever since W. T. Astbury published his first X-ray diffraction photographs we all knew that DNA had a crystalline structure, but until
the days of Crick and Watson no one knew what it was. The gap was
visible then, and if research had stopped in 1950 it would be visible
still; our picture of the world would be known to be imperfect. The
importance of Griffith's discovery was historical only (I do not mean
this in a depreciatory sense). He might not have made it; it might
not have been made to this very day; but if he had not, then some
other, different discovery would have served an equivalent purpose,
that is, would in due course have given away the genetic function
of DNA. The discovery of the structure of DNA was logically necessary for the further advance of molecular genetics. If Watson and
Crick had not made it, someone else would certainly have done so –
almost certainly Linus Pauling, and almost certainly very soon. It would
have been that same discovery, too; nothing else could take its place.

Watson and Crick (so Watson tells us) were extremely anxious that Pauling should *not* be the first to get there. In one uneasy hour they feared he had done so, but to their very great relief his solution was erroneous, and they celebrated his failure with a toast. Such an admission will shock most laymen: so much, they will feel, for the 'objectivity' of science; so much for all that fine talk about the disinterested search for truth. In my opinion the idea that scientists ought to be indifferent to matters of priority is simply humbug. Scientists are entitled to be proud of their accomplishments, and what accomplishments can they call 'theirs' except the things they have done or thought of first? People who criticise scientists for wanting to enjoy the satisfaction of intellectual ownership are confusing possessiveness with pride of possession. Meanness, secretiveness and sharp practice are as much despised by scientists as by other decent people in the world of ordinary everyday affairs; nor, in my experience, is generosity less common among them, or less highly esteemed.

It could be said of Watson that, for a man so cheerfully conscious of matters of priority, he is not very generous to his predecessors. The mention of Astbury is perfunctory and of Avery a little condescending. Fred Griffith is not mentioned at all. Yet a paragraph or two would have done it without derogating at all from the splendour of his own achievement. Why did he not make the effort?

It was not lack of generosity, I suggest, but stark insensibility. These matters belong to scientific history, and the history of science bores most scientists stiff. A great many highly creative scientists (I classify Jim Watson among them) take it quite for granted, though they are usually too polite or too ashamed to say so, that an interest in the history of science is a sign of failing or unawakened powers. It is not good enough to dismiss this as cultural barbarism, a coarse renunciation of one of the glories of humane learning. It points towards something distinctive about scientific learning, and instead of making faces about it we should try to find out why such an attitude is natural and understandable. A scientist's present thoughts and actions are of necessity shaped by what others have done and thought before him; they are the wave-front of a continuous secular process in which The Past does not have a dignified independent existence on its own. Scientific understanding is the integral of a curve of learning; science therefore in some sense comprehends its history within itself. No Fred, no Jim: that is obvious, at least to scientists; and being obvious it is understandable that it should be left unsaid. (I am speaking, of course, about the history of scientific endeavours and accomplishments, not about the history of scientific ideas. Nor do I suggest that the history of science may not be profoundly interesting as history. What I am saying is that it does not often interest the scientist as science.)

Jim Watson ('James' doesn't suit him) majored in zoology in Chicago and took his Ph.D. in Indiana, aged twenty-two. When he arrived in Cambridge in 1951 there could have been nothing much to distinguish him from any other American 'postdoctoral' in search of experience abroad. By 1953 he was world-famous. How much did he owe to luck?

The part played by luck in scientific discovery is greatly over-rated. *Ces hasards ne sont que pour ceux qui jouent bien*, as the saying goes. The paradigm of all lucky accidents in science is the discovery of penicillin – the spore floating in through the window, the exposed culture plate, the halo of bacterial inhibition around the spot on which it fell. What people forget is that Fleming had been *looking* for penicillin, or something like it, since the middle of the First World War. Phenomena such as these will not be appreciated, may not be knowingly observed, except against a background of prior expectations. A good scientist is discovery-prone. (As it happens there *was* an element of blind luck in the discovery of penicillin, though it was unknown to Fleming. Most antibiotics – hundreds are now known – are murderously toxic, because they arrest the growth of bacteria by interfering with metabolic processes of a kind that bacteria have in common with higher organisms. Penicillin is comparatively innocuous because it happens to interfere with a synthetic process peculiar to bacteria, namely the synthesis of a distinctive structural element of the bacterial cell wall.)

I do not think Watson was lucky except in the trite sense in which we are all lucky or unlucky – that there were several branching points in his career at which he might easily have gone off in a direction other than the one he took. At such moments the reasons that steer us one way or another are often trivial or ill thought-out. In England a schoolboy of Watson's precocity and style of genius would probably have been steered towards literary studies. It just so happens that during the 1950s, the first great age of molecular biology, the English Schools of Oxford and particularly of Cambridge produced more than a score of graduates of quite outstanding ability – much more brilliant, inventive, articulate and dialectically skilful than most young scientists; right up in the Watson class. But Watson had one towering advantage over all of them: in addition to being extremely clever he had something important to be clever *about*. This is an advantage which scientists enjoy over most other people engaged in intellectual pursuits, and they enjoy it at all levels of capability. To be a first-rate scientist it is not necessary (and certainly not sufficient) to be extremely clever, anyhow in a pyrotechnic sense. One of the great social revolutions brought about by scientific research has been the democratisation of learning. Anyone who combines strong common sense with an ordinary degree of imaginativeness can become a creative scientist, and a happy one besides, in so far as happiness depends upon being able to develop to the limit of one's abilities.

Lucky or not, Watson was a highly privileged young man. Throughout his formative years he worked first under and then with scientists of great distinction; there were no dark unfathomed laboratories in his career. Almost at once (and before he had done anything to deserve it) he entered the privileged inner circle of scientists among whom information is passed by a sort of beating of tom-toms, while others await the publication of a formal paper in a learned journal. But because it was unpremeditated we can count it to luck that Watson fell in with Francis Crick, who (whatever Watson may have intended) comes out in this book as the dominant figure, a man of very great intellectual powers. By all accounts, including Watson's, each provided the right kind of intellectual environment for the other. In no other form of serious creative activity is there anything equivalent to a collaboration between scientists, which is a subtle and complex business, and a triumph when it comes off, because the skill and performance of a team of equals can be more than the sum of individual capabilities. It was a relationship that did work, and in doing so brought them the utmost credit.

Considered as literature, *The Double Helix* will be classified under

Memoirs, Scientific. No other book known to me can be so described. It will be an enormous success, and deserves to be so – a classic in the sense that it will go on being read. As with all good memoirs, a fair amount of it consists of trivialities and idle chatter. Like all good memoirs it has not been emasculated by considerations of good taste. Many of the things Watson says about the people in his story will offend them, but his own artless candour excuses him, for he betrays in himself faults graver than those he professes to discern in others. *The Double Helix* is consistent in literary structure. Watson's gaze is always directed outward. There is no philosophising or psychologising to obscure our understanding; Watson displays but does not observe himself. Autobiographies, unlike all other works of literature, are part of their own subject-matter. Their lies, if any, are lies *of* their authors but not *about* their authors, who (when discovered in falsehood) merely reveal a truth about themselves, namely that they are liars. Although it sounds a bit too well remembered, Watson's scientific narrative strikes me as perfectly convincing. This is not to say that the apportionments of credits or demerits are necessarily accurate: that is something which cannot be decided in abstraction, but only after the people mentioned in the book have had their say, if they choose to have it. Nor will an intelligent reader suppose that Watson's judge-ments upon the character, motives and probity of other people (some-times apparently shrewd, sometimes obviously petty) are 'true' simply because he himself believes them to be so.

A good many people will read *The Double Helix* for the insight they hope it will bring them into the nature of the creative process in science. It may indeed become a standard case history of the so-called 'hypothetico-deductive' method at work. Hypothesis and inference, feedback and modified hypothesis, the rapid alternation of imaginative and critical episodes of thought – here it can all be seen in motion, and every scientist will recognise the same intellectual struc-ture in the research he does himself. It is characteristic of science at every level, and indeed of most exploratory or investigative processes in everyday life. No layman who reads this book with any kind of understanding will ever again think of the scientist as a man who cranks a machine of discovery. No beginner in science will henceforward believe that discovery is bound to come his way if only he practises a certain Method, goes through a certain well-defined performance of hand and mind.

Nor, I hope, will anyone go on believing that The Scientist is some definite kind of person. Given the context, one could not plausibly imagine a collection of people more different in origin and education, in manner, manners, appearance, style and worldly purposes than the men and women who are the characters in this book. Watson himself and Crick and Wilkins, the central figures; Dorothy Crowfoot and poor Rosalind Franklin, the only one of them not then living; Perutz, Kendrew and Huxley; Todd and Bragg, at that time holder of 'the most prestigious chair in science'; Pauling *père et fils*; Bawden and Pirie, in a momentary appearance; Chargaff; Luria; Mitchison and Griffith (John, not Fred) – they come out larger than life, perhaps, and as different one from another as Caterpillar and Mad Hatter. Watson's child-like vision makes them seem like the creatures of a Wonderland, all at a strange contentious noisy tea-party which made room for him because for people like him, at this particular kind of party, there is always room.

POSTSCRIPT

'Lucky Jim' was a defence of Watson against the storm of outraged criticism that burst out after the publication of The Double Helix. Nothing has occurred to shake my belief that the discovery of the structure and biological functions of the nucleic acids is the greatest achievement of science in the twentieth century. In defending Watson I felt much as advocates must feel when defending a client who is unmistakably guilty of many of the charges brought against him: I have in mind particularly his lack of adequate acknowledgement of the work of scientists such as Chargaff who made really important contributions to the elucidation of the problem which he and Crick finally solved. I showed Francis Crick my review before it appeared in the New York Review of Books and was very pleased when he said of my parallel with Alice's Wonderland and the Mad Hatter's tea-party: 'That's quite right, you know, it was exactly like that.'

One passage in this review of The Double Helix came in for a lot of criticism. I see it struck W. H. Auden[1] too, unfavourably I should guess. The passage (p. 275 above) runs:

It just so happens that during the 1950s, the first great age of molecular biology, the English schools of Oxford and particularly of Cambridge produced more than a score of graduates of quite outstanding ability – much more brilliant, inventive, articulate and dialectically skilful than most young scientists; right up in the Jim Watson class. But Watson had one towering advantage over all of them: in addition to being extremely clever he had something important to be clever *about*.

Surely, I was asked, you don't intend to imply that Shakespeare and Tolstoy etc. are not important and that it is hardly possible to be clever about them? Of *course* this is not what I intended. I had it in mind that many of the brilliant contemporaries of Jim Watson and many of the brightest literary students of the later 1950s entered the advertising or entertainment industries or contented themselves with petty literary pursuits. The widely prevalent opinion that almost any literary work, even if it amounts to no more than writing advertising copy or a book review, not to mention that Ph.D. thesis on 'Some little known laundry bills of George Moore', is intrinsically superior to almost any scientific activity is not one with which a scientist can be expected to sympathise.

[1] W. H. Auden, *A Certain World* (London, 1971).

HOUSE IN ORDER

ONE of the *New Yorker*'s happier literary inventions was the 'profile', a short non-partisan and as a rule absorbingly interesting essay on a person, a phenomenon or an idea. *Only One Earth*,[1] written for the United Nations Conference on Man and the Environment held in 1972, is a superbly accomplished profile of the world and of the sciences and technology that are responsible for its present delicate condition. Nearly seven pages are occupied by a list of the consultants to whom parts or the whole of the book were referred. This may give the impression that the book is essentially a pastiche or compendium of scholarly aphorisms. Nothing could be further from the truth – it is unmistakably the work of one mind, Barbara Ward Jackson's, with the name of René Dubos as an additional guarantee of scientific authenticity and of total integrity.

Only One Earth demonstrates how it is possible for someone of really first-rate intelligence to master the gist or general sense of many scientific ideas without having (something that would be impossible anyway) a technical knowledge of all the -ologies at bench or shop-floor level. I have read no work of popular science, nor indeed any work of any kind, that conveys so skilfully and justly a notion of the beauty, even majesty, of scientific ideas, and of the growth of scientific understanding.

Only One Earth gives the impression that scientific understanding, considered in the round, represents a superb and inspiring accomplishment of which human beings have every right to be exultantly proud. Moreover we are left in no doubt that it is towards science and an enlightened technology founded upon it that we must turn to find a remedy for our present condition. But that will hardly be enough unless we contrive to develop for the earth as a whole the deep and passionate sense of allegiance which youngsters are brought up to feel for their birthplace, school or nation. 'The planet is not yet a centre of rational loyalty for all mankind':

[1] Barbara Ward and René Dubos (London, 1972).

it is precisely this shift of loyalty that a profound and deepening sense of our shared and interdependent biosphere can stir to life in us. That man can experience such transformations is not in doubt. From family to clan, from clan to nation, from nation to federation, such enlargements of allegiance have occurred without wiping out the earlier loves. Today, in human society, we can perhaps hope to survive in all our prized diversity provided we can achieve an ultimate loyalty to our single, beautiful and vulnerable planet earth.

Only a tiny fraction of this book is in this elevated style. There is plenty of gritty material in it as well. We are all the better for learning, for example, that a new model incinerator built in Düsseldorf serves 700,000 people and yields revenues of $3.40 per ton of unprocessed refuse. The tone of the book is neither doomladen nor unreasonably sanguine, but it is, as it should be, extremely grave. Its message – which I believe to be true and fully justified by the text – is that we are at a pivotal point in history.

The two worlds of man – the biosphere of his inheritance, the technosphere of his creation – are out of balance, indeed potentially in deep conflict. And man is in the middle. This is the hinge of history at which we stand, the door of the future opening onto a crisis more sudden, more global, more inescapable and more bewildering than any ever encountered by the human species and one which will take decisive shape within the life span of children who are already born.

*

The message of John Maddox's book[1] may be summed up by saying that he believes that this is all greatly exaggerated. John Maddox as Editor of *Nature* holds the most prestigious post available to a science journalist in the UK. He is a widely learned man and an old hand at scientific journalism. The following passage shows what a good journalist he is: 'It used to be commonplace for men to parade the city streets with sandwich boards proclaiming "The end of the world is at hand!" They have been replaced by a throng of sober people, scientists, philosophers, politicians, proclaiming that there are other more subtle calamities just around the corner.'

This is very good writing of its kind: it is witty and sharply pointed, and the point he makes could hardly have been made more briefly. On the other hand, chapter headings like 'The numbers game' (popu-

[1] John Maddox, *The Doomsday Syndrome* (London, 1972).

lation projection) and 'Ecology is a state of mind' show how mischievous journalism of this kind can be.

The work which typifies the state of mind to which *The Doomsday Syndrome* professes to be the antidote is the issue for January 1972 of the *Ecologist* – the issue bearing the brave title 'Blueprint for Survival'; the principal contention of 'Blueprint' is that the human race cannot sustain an ethos of continuous and indefinite expansion. It is indeed a policy that has already got us into a precarious mess. A small fraction of the world can continue to improve its standard of living, but only at the expense of degrading or failing to improve the lot of the remainder. In spite of their lesser numbers, the highly industrialised fraction of the world consumes 80 per cent of the world's energy and raw materials. Although the populations of both worlds are increasing at a dangerous rate, the growth of the industrialised moiety represents by far the greater danger. The world cannot sustain indefinitely the demands at present being made upon it: resources are finite and the waste products of civilisation's metabolism are reaching dangerous levels. Worse still, the cycles of nature are being disrupted so that natural regenerative processes are impeded or stopped altogether.

Maddox bustles in with the assurance that most of this is gravely overstated. Yet 'Blueprint for Survival' had a profound influence in England and elsewhere partly because its thesis was endorsed (with unspecified reservations) by upwards of a dozen assorted scientists, economists or public figures. Although it could be (and was) objected that very few of them had any expert knowledge of the matters under discussion, it could be said in defence that the signatories were united by a fearful anxiety (which Maddox would say was groundless) about the future of the planet. It is easy to go from the position that the extinction of a few species of songbirds is not a world-wide catastrophe, and that the inconvenience caused to certain sea-birds by the perilous thinness of the shells of their eggs (the effects of DDT) is something we must simply learn to bear with fortitude, to taking the view that overpopulation as a calamity that is mainly going to affect Blacks and other coloured people who must learn to expiate their exuberant fecundity by nature's own penances: death by starvation and infectious disease. I do not for one moment impute any such insensibility to John Maddox, and can easily understand why he is infuriated by the assumption of many ecologists that they are the keepers of the world's conscience while all others pursue the goal of self-interest with callous indifference to the fate of the planet. Yet it is a matter of historical fact that all great reforming movements, including those which in retrospect we see to have been well founded and to have had a salutary effect, have been distinguished by a sense of essential rightness and the conviction that those who disagree with them belong in outer darkness. It is this conviction that provides any such movement with its impetus, and attracts disciples. Let us study Maddox's counter-arguments in a little detail.

To some extent he is certainly justified. To describe over-population as an important cause of national conflict goes against all historical evidence. Is a state of warfare so very much more common nowadays than it was in the meagrely populated Europe of the Thirty Years War? It is also fair to ask, as Maddox does, if the people of the Netherlands, one of the most overcrowded of nations, are more aggressive than the people of the US. Maddox is very scornful about inferences relating to the behaviour of human beings drawn from the study of laboratory animals, particularly rats and mice. But Dear God, almost the whole of modern medical science is founded upon observations made in the first place upon laboratory animals. Without them there would be no penicillin, no insulin, no transplantation, and almost certainly no John Maddox either. The Editor of *Nature* should know better than to write like this. In any case, he would be far better occupied in exploding some of the myths of population lore – the great suicide march of the lemmings, for example, a phenomenon of the very highest degree of implausibility, or the notion that population numbers can be stabilised by fiscal or didactic procedures.

Another argument of Maddox's which ought not to command much sympathy is his hint that the movement against over-population represents a resurgence of radical puritanism. 'Is it possible', he asks, 'that some of the emotion with which the issue of population growth has been invested, stems from a sneaking suspicion that procreation is a sinful business?' This is neither a fair nor a sensible question, because it could be refuted only by showing that population experts exult in procreation. Nevertheless, Maddox takes his idea very seriously: 'this said,' he goes on to say, 'it is no surprise that the growth of population is central to the Doomsday movement'. Maddox could have made a better case against population forecasting by pointing out that as recently as 1945 the world's greatest doomographer, Dr Alfred J. Lotka, wrote a technical paper sadly bewailing what he believed to be the inescapable truth that the peoples of the Western world were going to die out through infertility, as so many people believed in the 1930s. Population growth depends upon behavioural, social and political variables the values of which cannot be accurately predicted. However, the population case does not really rest upon projection into the distant future, though some people give the mistaken impression that the Malthusian apocalypse won't really be upon us until the turn of the millennium. The fact is that the opening bars of the *Dies irae* are sounding right now, inasmuch as death by starvation and infectious disease is already commonplace in over-populated parts of the world. Moreover there is precious little sign of any modification or abatement of the forces which have brought these parts of the world into their present condition.

I learned with interest and surprise from Maddox that a reading of Malthus's essay on population induced William Pitt to withdraw his Poor Law. As the Poor Law was obviously a good thing, this might induce us to believe, quite wrongly, that a modern legislature would be equally at fault if thoughts of population growth induced it to try to withdraw whatever fiscal inducements still remain for married couples to have large families. Such an inference will be totally unwarranted, of course, which is by no means to assert that demographic factors are of overriding importance in deciding upon the propriety of family allowances, income tax rebates, and other such financial encouragements of fertility.

I think Maddox is hopelessly sanguine in supposing that improvement in the standard of living that accompanies industrialisation will in itself lower the fertility of under-developed peoples. My own long-term hope for the future, as a biologist, is that the prolonged and universal practice of family limitation will, by removing the selective differential in favour of families of high fertility, eventually bring about a general lowering of the fertility level of human beings. This would bring the human species into line with all those other species in which the level of innate fertility or fecundity is very nicely adjusted to their chances of survival. It is a most unfortunate thing that most school-children who could 'do' biology were introduced to the concepts of Darwinism by a particularly misleading syllogism which runs as follows:

1 All organisms produce offspring in numbers vastly in excess of their requirements.

2 Only a minority of these survive.

3 Therefore only those animals survive which are best fitted to do so.

The error here lies in the major premise that organisms produce offspring in numbers vastly in excess of their needs. Actually, as Fisher long ago pointed out, fecundity is itself subject to natural selection, and the numbers of offspring produced by an animal are closely related to the forces of mortality it is subjected to, so that most species produce just about the number of young sufficient and necessary to perpetuate their kind.

Maddox is on pretty solid ground when he sweeps aside the alleged dangers of 'genetic engineering', pointing out that artificial selection, which has been practised for centuries, is, in effect, a form of 'genetic engineering'. The injudicious procedures which have given us so many bizarre breeds of dog and so many hideous cultivated flowers are not, however, what the avant-garde means by 'genetic engineering', which refers rather to a direct intervention with the germinal DNA or to procedures such as test-tube fertilisation of human ova followed by implantation into the womb – a basically inoffensive procedure which is not different in genetic principle from adoption. The more extravagant claims for genetic engineering, however, anticipate that it will run to a direct repair of germinal DNA if it is in any respect faulty. Most of these claims are regarded by a number of all-round biologists such as MacFarlane Burnet, Jacques Monod and myself as extremely far-fetched. But in any case genetic engineering does not pose the *kind* of threat that makes people really anxious. It is indeed possible in principle to 'clone' human egg cells and reimplant them so as to produce very numerous replicas of some single chosen human being. But God has put it into the power of every one of us to refrain from doing anything of the kind. Thus the entire procedure has not the awful air of inevitability about it which seems to distinguish some of the real threats facing mankind. The same considerations apply to the transplantation of the head. This transplantation is for all practicable purposes impossible because a transplanted head will not make nervous connections with the body of which it should be the administrative centre. But people are not *really* frightened about the creation of races of two-headed men, because their instinct tells them how easy it is *not* to transplant the head, and how easily the ambition to do so could be thwarted.

What really upsets people is the feeling that over-population and pollution and other such hazards follow inevitably from our existing style of life. But the fears aroused by man-made monsters and by pollution are of an entirely different kind, and Maddox's power to allay the one does not imply that he has done anything to allay the other.

Maddox sees the present pollution scare as a carry-over from our genuinely well-founded fears of radioactive pollution in the 1950s, and there is no attempt to depreciate the malignant effects of radioactive pollution. But what, we may ask, would the levels of radioactive pollution be today if large numbers of, as some people thought them at the time, tiresome and scary agitators had not made a tremendous fuss about the hazards of radioactivity in the '50s? Now that the dangers are fully recognised and pretty well assessed, it begins to be possible to have some confidence in the skill of engineers and the sanity of administrators to prevent the worst consequences of radioactive contamination as they were at one time feared. The public indignation that led to the institution of committees of surveillance under the Medical Research Council in the UK and the National Academy of Sciences in the USA represent a very proper democratic reaction and a natural expression of a citizen's dislike of being pushed around. History is now repeating itself in a cognate context.

Maddox on pollution is good, punchy pamphleteering prose, but while it may be true that the reports that Lake Erie is dead or on its deathbed have been greatly exaggerated, what about the former plight of Lake Vättern? The Swedes are practical-minded and sensible people, and their Government has not spent tens of thousands of pounds on the reclamation of Lake Vättern in order to capture what might be called the apocalyptic vote. Nor have policemen on traffic duty in Tokyo worn gas-masks to amuse the children. It seems to me that Maddox's attack on the views of Rachel Carson is directed not so much against what she actually said as against a certain conventional misrepresentation of her views. Maddox's conclusion on the matter is that 'ecology implies not that pesticides are too dangerous to use but that they should be used cleverly'. This was precisely Miss Carson's view. Just as, so far from having to wait for the turn of the millennium, we can see the malignant effects of over-population today, so also can we already see the malignant effects of pollution. A railway journey south of New York or north of Birmingham in England can give one the impression of visiting a battlefield – the scene of some tremendous victory of technology over nature. It is in these wastelands that the spring is silent. Miss Carson's book started a movement of questioning and vigilance which is wholly admirable. If it is a bit overstated in places, why then one sometimes has to raise one's voice to make it heard.

One does not have to be a classical scholar to know that the words 'ecumenics', 'ecology' and 'economics' have all the same root in the Greek word *oikos*, meaning 'house'. It is the great strength of Barbara Ward's text that she argues so cogently for the necessity of putting and keeping our house in order, and it is the weakness of Maddox's text that he does not seem clearly enough to realise that there is anything very serious to worry about. The great message of the Japanese to the United Nations Conference might be summarised as 'For God's sake don't make the kind of mistakes that we have made.' Perhaps they should invite Mr Maddox over to reassure them.

A BIOLOGICAL RETROSPECT

THE title of my presidential address,[1] you will have discerned, is 'A Biological Retrospect', and on the whole it has not been well received. 'Why a biological *retrospect*?', I have been asked; would it not be more in keeping with the spirit of the occasion if I were to speak of the future of biology rather than of its past? It would indeed be, if only it were possible, but unfortunately it is not. What we want to know about the science of the future is the content and character of future scientific theories and ideas. Unfortunately, it is impossible to predict new ideas – the ideas people are going to have in ten years' or in ten minutes' time – and we are caught in a logical paradox the moment we try to do so. For to predict an idea is to have an idea, and if we have an idea it can no longer be the subject of a prediction. Try completing the sentence 'I predict that at the next meeting of the British Association someone will propound the following new theory of the relationships of elementary particles, *namely* . . .'. If I complete the sentence, the theory will not be new next year; if I fail, then I am not making a prediction.[2]

Most people feel more confident in denying that certain things will not come to pass than in declaring that they can happen or surely will happen. Many a golden opportunity to remain silent has been squandered by anti-prophets who do not realise that the grounds for declaring something impossible or inconceivable may be undermined by new ideas that cannot be foreseen. Here is an instructive passage from the philosophic writings of a great British physiologist, J. S. Haldane (father of J.B.S.). It comes from *The Philosophy of a Biologist* of 1931, and its subject is the nature of memory, in a very general sense that includes 'genetic memory' – for example, the faculty or endowment which ensures that a frog's egg develops into a frog and, indeed, into a frog of a particular kind.

Haldane is very critical of the theories of memory propounded by

---

[1] To the Zoological Section of the British Association for the Advancement of Science, 1965.

[2] See 'Expectation and Prediction', pp. 298–310 below.

288   A BIOLOGICAL RETROSPECT

Ewald Hering and Richard Semon, who 'assume that memory in general is dependent on protoplasmic "engrams", and that germ-cells are furnished with a system of engrams, functioning as guide-posts to all the normal stages of development'. ('Engrams', I should explain, are more or less permanent physical memory traces or memory imprints that are thought to act as directive agencies in development.)[1] 'This theory', Haldane goes on to say, 'has quite evidently all the defects of other attempts at mechanistic explanations of development. How such an amazingly complicated system of signposts could function by any physico-chemical process or reproduce itself indefinitely often is inconceivable.'[2]

What Haldane found himself unable even to conceive is today a commonplace. Only twelve years after the publication of the passage I quote, Avery and his colleagues had determined the class of chemical compound to which genetic engrams belong. In the meantime our entire conception of 'the gene' has undergone a revolution. Genes are not, as at one time or another people have thought them, samples or models; they are not enzymes or hormones or prosthetic groups or catalysts or, in the ordinary sense, agents of any kind. Genes are *messages*. I think Hans Kalmus[3] was the first to use this form of words, but the idea that a chromosome is a molecular code script containing a specification of development is Schrödinger's.[4]

My purpose in this address is to identify some of the great conceptual advances that have taken place during the past twenty-five or thirty years on four different planes of biological analysis. Working biologists, I should explain, tend nowadays to classify themselves less by 'subjects' than by the analytical levels at which they work – a horizontal classification where the older was vertical. So we have molecular biologists, whose ambition is to interpret biological performances explicitly in terms of molecular structure; we have cellular biologists,

[1] See R. Semon, *The Mneme* (London, 1921), particularly pp. 24, 113, 180, 211; and Hering's paper 'On Memory, a Universal Attribute of Organized Matter' in *Almanach der kaiserlichen Akademie der Wissenschaften in Wien*, **20**, 253, 1870.
[2] *The Philosophy of a Biologist* (London, 1931), p. 162.
[3] 'A Cybernetical Aspect of Genetics', *Journal of Heredity*, **41**, 19, 1950.
[4] E. Schrödinger, *What is Life?* (Cambridge, 1944), especially pp. 19–20, 61–2, 68. For Weismann's far-sighted views on the matter, see S. Toulmin and J. Goodfield, *The Architecture of Matter* (London, 1962).

biologists who work at the level of whole organisms (the domain of classical physiology), and biologists who study communities or societies of organisms. We can discern each of these four strata within each 'subject' of the traditional, that is, the vertical, classification. There are molecular and cellular geneticists, geneticists in Mendel's sense, and population geneticists. So also in endocrinology or immunology: each is now studied at the molecular and cellular level as well as at the level of whole organisms. They abut into the population level, too: we study the effects of crowding and fighting on the adrenals and so indirectly on reproductive performance, and we study the epidemiological consequences of natural or artificial immunisation and the evolutionary consequences of epidemics. I have noticed that a biologist's interest and understanding, and also, in a curious way, his loyalties, tend to spread horizontally, along strata, rather than up and down. Our instinct is to try to master what belongs to our chosen plane of analysis and to leave to others the research that belongs above that level or below. An ecologist in the modern style, a man working to understand the agencies that govern the structure of natural populations in space and time, needs much more than a knowledge of natural history and a map. He must have a good understanding of population genetics and population dynamics generally, and certainly of animal behaviour; more than that, he must grasp climatic physiology and have a feeling for whatever may concern him among the other conventional disciplines in biology (I have already mentioned immunology and endocrinology). There is no compelling reason why he should be able to talk with relaxed fluency about messenger-RNA, and it is not essential that he should ever have heard of it – though an unreasonable feeling that he 'ought' to know something about it is more likely to be found in a good ecologist than in an indifferent one.

I shall now take one example from each of these four planes of biological analysis and try to show how our ideas have changed since the last Cambridge meeting of the British Association in 1938 – a period that corresponds roughly with my own professional lifetime.

Population genetics and evolution theory

Biologists of my generation were still brought up in what I call the 'dynastic' concept of evolution. The course of evolution was unfolded to us in the form of pedigrees or family trees, and we used the old language of universals in speaking of the evolution of *the* dogfish, *the* horse, *the* elephant, and, needless to say, of Man.

The dynastic conception coloured our thoughts long after the revival of Darwinism had made it altogether inappropriate. By the 'revival of Darwinism' I mean the reformulation of Darwinism in the language of Mendelian genetics – the work, as we all so very well know, of Fisher, J. B. S. Haldane, Wright, Norton, and, in a rather qualified sense, of Lotka and Volterra. The subject of evolutionary change, we now learned, was a population, not a lineage or pedigree: evolution was a systematic secular change in the genetical structure of a population, and natural selection was overwhelmingly its most important agent. But to those brought up in the dynastic style of thinking about evolution it seemed only natural to suppose that the outcome of an evolutionary episode was the devising of a new geno-type – of that new genetical formula which conferred the greatest degree of adaptedness in the prevailing circumstances. This improved genetic formula – this new solution of the problem of remaining alive in a hostile environment – would be shared by the great majority of the members of the population, and would be stable except in so far as it might be modified by further evolution. The members of the population were predominantly uniform and homozygous in genetic make-up, and, to whatever degree they were so, would necessarily breed true. Genetic diversity was maintained by an undercurrent of mutation, but most mutants upset the hard-won formula for adapted-ness, and natural selection forced them into recessive expression, where they could do little harm. When evolution was not in progress natural selection made on the whole for uniformity. Polymorphism, the occurrence of a stable pattern of genetic inequality, was recognised as an interesting but somewhat unusual phenomenon, each example of which required a special explanation, that is, an explanation peculiar to itself.

These ideas have now been superseded, mainly through the empirical discovery that naturally outbreeding populations are highly diverse. Chemical polymorphism (allotypy)[1] is found wherever it is looked

---

[1] A term coined by J. Oudin to describe gamma globulin variants: it might well be generalised to include all molecular polymorphism.

for intently enough by methods competent to reveal it. The molecular variants known in human blood alone provide combinations that far outnumber the human race – variants of haemoglobin, non-haemoglobin proteins, and red-cell enzymes; of red-cell antigens and white-cell antigens; and of haptoglobins, transferrins and immunoglobulins. Today it is no longer possible to think of the evolutionary process as the formulation of a new genotype or the inauguration of a new type of organism enjoying the possession of that formula. The *product* of evolution is itself a population – a population with a certain newly devised and well-adapted pattern of genetic *in*equality. This pattern of genetic differentiation is shaped and actively maintained by selective forces: it is the population as a whole that breeds true, not its individual members. We can no longer draw a distinction between an active process of evolution and a more or less stationary end-product: evolution is constantly in progress, and the genetical structure of a population is actively, that is dynamically, sustained.

These newer ideas have important practical consequences. The older outlook was embodied in that older, almost immemorial ambition of the livestock breeder, to produce by artificial selection a true breeding stock with uniform, and uniformly desirable, characteristics; and this was also the ambition – sometimes kindly, but always mistaken – of old-fashioned 'positive' eugenics. It now seems doubtful if, with free-living and naturally out-breeding organisms, such a goal can ever be achieved. Modern stock-breeders tend to adopt a very nicely calculated regimen of cross-breeding which, abandoning the goal of a single self-perpetuating stock, achieves a uniform marketable product of hybrid composition. The genetical theory underlying this scheme of breeding embodies, and was indeed partly responsible for, the newer ideas of population structure I have just outlined.

I cannot predict what new ideas will illuminate the theory of evolution in future, but it is not difficult to guess the contexts of thought in which they are likely to appear. The main weakness of modern evolutionary theory is its lack of a fully worked out theory of variation, that is, of *candidature* for evolution, of the forms in which genetic variants are proffered for selection. We have therefore no convincing account of evolutionary progress – of the otherwise inexplicable tendency of organisms to adopt ever more complicated solutions of the problems of remaining alive. This is a 'molecular' problem, in the newer biological usage of that word, because its working out depends on a deeper understanding of how the physico-chemical properties and behaviour of chromosomes and nucleo-proteins generally qualify them to enrich the candidature for evolution; and this reflection is my cue to turn to conceptual advances in biology at the molecular level.

The physical basis of life

In the early 1930s no one knew what to make of the nucleic acids. Bawden and Pirie had not yet shown that nucleic acid was an integral part of the structure of tobacco mosaic virus, and we were still a decade from the astonishing discovery by Avery and his colleagues, in the Rockefeller Institute, that the agent responsible for pneumococcal transformations was a deoxyribonucleic acid.

Since there was nothing very much to say about nucleic acids you may well wonder what everybody *did* talk about. One topic of conversation was the crystallisation of enzymes. Sumner had crystallised urease in 1926 and Northrop pepsin in 1930; soon Stanley would crystallise tobacco mosaic virus, at that time still thought to be a pure protein. But the most exciting and, as it seemed to us, portentous discoveries were those of W. T. Astbury, whose X-ray diffraction pictures of silk fibroin and of hair and feather keratins had revealed an essentially crystalline orderliness in ordinary biological structures. For some purposes, however, X-ray analysis was too powerful. The occasion called for resolving powers between those of the optical microscope and the X-ray tube, and this need was fulfilled by electron microscopy. I saw my first electron-photomicrograph in *Nature* in 1933; its resolving power was then one micron.

Electron microscopy has shown that cells contain sheets, tubes, bags and, indeed, micro-organs – real anatomical structures in the sense that they have firm and definite shapes and look as if only their size prevented our picking them up and handling them. Moreover, there is no dividing line between structures in the molecular and in the anatomical sense: macromolecules have structures in a sense intelligible to the anatomist and small anatomical structures are molecular in a sense intelligible to the chemist. (Intelligible *now*, I should add: as

Pirie[1] has told us, the idea that molecules have literally, that is, spatially, a structure was resisted by orthodox chemists, and the credentials of molecules with weights above 5,000 were long in doubt.) In short, the orderliness of cells is a structured or crystalline orderliness – a 'solid' orderliness, indeed, for 'the so-called amorphous solids are either not really amorphous or not really solid'.

This newer conception represents a genuine upheaval of biological thought, and it marks the disappearance of what may be called the *colloidal* conception of vital organisation, itself a sophisticated variant of the older doctrine of 'protoplasm'. The idea of protoplasm as a fragile colloidal slime, a sort of biological ether permeating otherwise inanimate structures, was already obsolete in the '30s; even then no one could profess to be studying 'protoplasm' without being thought facetious or slightly mad. But we still clung to the colloidal conception in its more sophisticated versions, which allowed for heterogeneity and for the existence of liquid crystalline states, and it was still possible to applaud Hopkins's famous aphorism from the British Association meeting of 1913, that the life of the cell is 'the expression of a particular dynamic equilibrium in a polyphasic system'. For, inadequate though the colloidal conception was seen to be, there was nothing to take its place. Peters's idea of the existence of a 'cytoskeleton' to account for the orderly unfolding of cellular metabolism in time and place now seems wonderfully prescient, but there was precious little direct evidence for the existence of anything of the kind, and much that seemed incompatible with it.

The substitution of the structural for the colloidal conception of 'the physical basis of life' was one of the great revolutions of modern biology; but it was a quiet revolution, for no one opposed it, and for that reason, I suppose, no one thought to read a funeral oration over protoplasm itself.

Cellular differentiation in embryonic development

Embryology is in some ways a model science. It has always been distinguished by the exactitude, even punctilio, of its anatomical descriptions. An experiment by one of the grand masters of embryology

[1] N. W. Pirie, 'Patterns of Assumption about Large Molecules', *Archives of Biochemistry and Biophysics*, supp. 1, 21, 1962.

could be made the text of a discourse on scientific method. But something is wrong; or has been wrong. There is no *theory* of development, in the sense in which Mendelism is a theory that accounts for the results of breeding experiments. There has therefore been little sense of progression or timeliness about embryological research. Of many papers delivered at embryological meetings, however good they may be in themselves – they are sometimes marvels of analysis, and complete and satisfying within their own limits – one too often feels that they might have been delivered five years beforehand without making anyone much the wiser, or deferred for five years without making anyone conscious of great loss.

It was not always so. In the 1930s experimental embryology had much the same appeal as molecular biology has today: students felt it to be the most rapidly advancing front of biological research. This was partly due to the work of Vogt, who had shown that the mobilisation and deployment of cellular envelopes, tubes and sheets was the fundamental stratagem of early vertebrate development (thus re-laying the foundations of comparative vertebrate embryology); but it was mainly due to the 'organiser theory' of Hans Spemann, the theory that differentiation in development is the outcome of an orderly sequence of specific inductive stimuli. The underlying assumption of the theory (though not then so expressed) was that we should look to the chemical properties of the inductive agent to find out why the amino-acid sequence of one enzyme or organ-specific protein should differ from the amino-acid sequence of another. The reactive capabilities of the responding tissue were emphasised repeatedly, but only at a theoretical level, for 'competence' did not lend itself to experimental analysis, and the centre of gravity of actual research lay in the chemical definition of inductive agents.

Wise after the event, we can now see that embryology simply did not have, and could not have created, the background of genetical reasoning which would have made it possible to formulate a theory of development. It is not now generally believed that a stimulus external to the system on which it acts can specify the primary structure of a protein, that is, convey instructions that amino-acids shall be assembled in a given order. The 'instructive' stimulus has gone the way of the philosopher's stone, an agent dimly akin to it in certain ways. Embryonic development at the level of molecular differentiation must therefore be an unfolding of pre-existing capabilities, an acting-out of genetically encoded instructions; the inductive stimulus is the agent that selects or activates one set of instructions rather than another. It is just possible to see how something of the kind happens in the induction of adaptive enzymes in bacteria – a phenomenon of which the older description, the 'training' of bacteria, reminds us that it too, at one time, was thought to be 'instructive' in nature. All this applies only to biological order at the level of the amino-acid sequences of proteins or the nucleotide sequences of nucleid acids. Nothing is yet known about the genetic specification of order at levels above the molecular level.

The function performed by the hierarchy of inductive stimuli as it occurs in vertebrate development is to determine the specificities of time and place: it is an inductive stimulus which determines that a lens shall form here and not there, now and not then. As I see it, it is the inductive process that allows vertebrate eggs and embryos before gastrulation to indulge in the prodigious range of adaptive radiation to be seen in germs as disparate as a dogfish's egg and a human being's – a case I have argued elsewhere and need not go over here again.[1]

Biology of the organism: animal behaviour

If experimental embryology was the subject that seemed most exciting to students of the '30s, that most nearly on the threshold of a grand revelation, the study of animal behaviour (in the sense in which we now tend to use the word 'ethology') seemed just as clearly the most frustrating and unrewarding. Twenty years later it was the other way about: embryology had lost much of its fascination and many of the ablest students were recruited into research on behaviour instead. What had happened in the meantime?

In the early 1930s we had one new behavioural concept to ponder on: the idea that an animal might in some way apprehend a sensory pattern or a behavioural situation as a whole and not by a piecing together of its sensory or motor parts. That was the lesson of *Gestalt* theory. We had also learnt finally, and I hope for ever, the method-ological lesson of behaviourism: that statements about what an animal

[1] *Journal of Embryology and Experimental Morphology*, **2**, 172, 1954.

feels or is conscious of, and what its motives are, belong to an altogether different class from statements about what it does and how it acts. I say the 'methodological' lesson of behaviourism, because behaviourism also stands for a certain psychological theory, namely, that the phenomenology of behaviour is the whole of behaviour – a theory of which I shall only say that, in my opinion, it is not nearly as silly as it sounds. Even the methodology of behaviourism seemed cruelly austere to a generation not yet weaned from the doctrine of privileged insight through introspection. But what comparable revolution of thought ushered in the study of animal behaviour in the style of Lorenz and Tinbergen and led to the foundation of flourishing schools of behaviour in Oxford, in Cambridge, and throughout the world?

I believe the following extremely simple answer to be the right one. In the '30s it did not seem to us that there *was* any way of studying behaviour 'scientifically' except through some kind of experimental intervention – except by confronting the subject of our observations with a 'situation' or with a nicely contrived stimulus and then recording what the animal did. The situation would then be varied in some way that seemed appropriate, whereupon the animal's behaviour would also vary. Even poking an animal would surely be better than just looking at it: *that* would lead to anecdotalism; that was what bird-watchers did.

Yet it was also what the pioneers of ethology did. They studied natural behaviour instead of contrived behaviour, and were thus able for the first time to discern natural behaviour-structures or episodes – a style of analysis helped very greatly by the comparative approach, for the occurrence of the same or similar behavioural sequences in members of related species reinforced the idea that there was a certain natural connectedness between its various terms, as if they represented the playing out of a certain instinctual programme. Then, and only then, was it possible to start to obtain significant information from the study of contrived behaviour – from the application or withholding of stimuli – for it is not informative to study variations of behaviour unless we know beforehand the norm from which the variants depart.

The form of address I chose – to trace the recent growth and transformation of ideas in four 'subjects' belonging to four levels of biological analysis – gave me no opportunity to mention some of the greatest innovations of modern scientific thought: the dynamical state of bodily constituents, for example, the perpetual flux of the material ingredients of the body. Nor have I said anything except by implication of the greatest discoveries in modern science, those which revealed the genetical functions of the nucleic acids. Yet I feel I have said enough of the growth of biology in the recent past to draw some morals, however trite.

The history of animal behaviour – in particular the sterility of the older experimental approach – illustrates the danger of doing experiments in the Baconian style; that is to say, the danger of contriving 'experiences' intended merely to enlarge our general store of empirical knowledge rather than to sustain or confute a specific hypothesis or presupposition. The history of embryology shows the dangers of an imagined self-sufficiency, for embryology is an inviable fragment of knowledge without genetics. (I often wonder what academics mean when they say of a certain subject that it is a 'discipline in its own right'; for what science is entire of itself?) You may think our recent history entitles us to feel pretty pleased with ourselves. Perhaps: but then we felt pretty pleased with ourselves twenty-five years ago, and in twenty-five years' time people will look back on us and wonder at our obtuseness. However, if complacency is to be deplored, so also is humility. Humility is not a state of mind conducive to the advancement of learning. No one formula will satisfy that purpose, for there is no one kind of scientist; but a certain mixture of confidence and restless dissatisfaction will be an ingredient of most formulae. Confident we may surely be, for the next twenty-five years will throw up many new ideas as profound and astonishing as any I have yet described, namely . . . but I have no time left to tell you what they are.

EXPECTATION AND PREDICTION

WISE folk may or may not form expectations about what the future holds in store but the foolish can be relied upon to predict with complete confidence that certain things will come about in the future or that others will not.

It is well worth while insisting upon the clear distinction of meaning between the two. A prediction always pretends to foreknowledge where an expectation is merely a hypothesis with a future setting ('I expect *Pluto's Republic* to be on sale in October 1982') – a hypothesis which the passage of time will either corroborate or confound. We cannot be viable human beings without taking some view of what will happen in the future. We confidently expect that the sun will rise tomorrow morning: it has become a habit of thought, but as the great Scottish philosopher David Hume (1711–76) pointed out we should be plunged into a labyrinth of philosophical difficulties if we attempted to *prove* that the sun would rise tomorrow on the basis that such a declaration would have been true every yesterday.

Astronomical predictions are perhaps the most famous of all and have in the past been the most awe-inspiring. They extend all the way from predictions such as one finds in nautical almanacs, giving us the exact times of sunrise, sunset, phases of the moon, high tides etc.; but grander than all of these, in the year 1704 the English astronomer and mathematician Edmund Halley predicted that the comet which now bears his name would return in 1758 – it is again expected in this neck of the woods in 1986.

Surely such predictions embody foreknowledge? Not really: the future position of the comet is deduced from our *present* knowledge of the comet's whereabouts and of the nature of its orbit around the sun. Halley was in a specially good position to acquire this knowledge, for he was a friend of Isaac Newton's and thoroughly familiar with Newton's *Principia Mathematica*, widely thought of as the greatest of all scientific works, which he proof-read and saw through the press. Halley's spectacular display of apparent foreknowledge was based upon what he knew already of the comet's position plus a number of astronomical principles which he already knew, or thought or assumed he knew. If Halley had been mistaken it could only have been because the knowledge he thought he possessed in 1704 was mistaken, or his logical reasoning (that is, his mathematical calculation) was erroneous.

Astronomy and sociology: historicism

Halley's prediction made a tremendous impression and seemed to many people to be the very paradigm of all that is truly 'scientific'. It inflamed sociologists and historians with the ambition to devise a historical social science that would embody the laws of social transformation and of the historic process and thus make it possible to foretell the future of mankind – a truly scientific sociology, it was thought, would embody the laws of human destiny and so make historic prediction possible. The belief that a predictive science of history can exist is known as *historicism*.

To those familiar with such matters it is very well known that David Hume punctured the pretensions of the philosophic doctrine known as 'empiricism', the belief that upon the evidence of the senses – upon sensory information only – it is possible to propound scientific laws of apodictic certainty. Yet many who know of Hume's scepticism and the revolution he brought about in undermining empiricism are apparently unaware that the Vienna-born philosopher Karl Raimund Popper demolished historicism as effectively as Hume demolished empiricism.

Popper's argument turns upon the theorem that I shall attempt to derive formally below, namely that it is not possible to predict scientific ideas or advances in science. Popper's argument goes thus: the course of history is influenced – no matter how or how much – by advances of science and technology;[1] but advances in science and technology cannot be predicted: therefore the future course of history cannot be predicted.

---

[1] For example, it is generally admitted to be true that Germany would not have been able to embark upon the First World War unless Fritz Haber had worked out how to 'fix' atmospheric nitrogen and turn it into nitrates or compounds of ammonia, so making Germany independent of imported fertilisers. This is the best example I know of how advances in technology can influence the course of history.

An example of historicism: The economic interpretation of the process of history embodied in Marxism is the best-known form of historicist argument. I cite now an economic interpretation of history that illustrates very clearly the kind of fallacy to which Karl Popper called special attention. Consider the factors that determine the positioning, size and shape of factories and the whereabouts of the people who work in them. The source of energy being coal, which is costly to transport, factories must grow up alongside or near coalfields so as to reduce to the minimum the cost of transporting coal. The dwelling houses of workers will be near the factories, so a township must grow up in the neighbourhood of coalfields.

Coal is converted into usable energy through the medium of steam; it is not, however, feasible for each power-tool to be worked by its own little steam-engine. Factories will therefore be so arranged that a single big steam-engine can drive a single overhead shaft or spindle to which individual machine-tools are connected by flexible leather or rubber belts. Individual factories will get larger and larger so as to increase the number of machine-tools that can be served by a single steam-engine.

It all sounds perfectly plausible, but the argument is mistaken, for it is not true that factories get larger and larger and that they congregate nearer and nearer coalfields. Neither Marx nor any other devotee of the economic interpretation of history was able to foresee the coming of electrical power, one of the consequences of which was that factories no longer needed to be built near coalfields; moreover the properties of electrical power are such as to make possible the springing up of numerous small factories containing machine-tools individually powered by their own electrical outlets. This is just what has happened in the British Midlands and the State of Massachusetts – Marxists were wrong in the historical prediction based upon economic theory for the simple reason that Marx was unable to foresee the coming of electrical energy. A much more deeply erroneous prediction is that which turns upon the notion of class warfare. According to Marxist theory the direction of the flow of history is shaped by a struggle for supremacy between social classes – particularly between the proletariat and those who own the means of production. Marx predicted that the struggle between the classes would inevitably lead to a social revolution followed by the victory of the proletariat and the disappearance of class stratification. Because of the paramount necessity to maximise profits, the rate of pay, degree of freedom and general welfare of the working class must inevitably deteriorate – and this would be a principal factor in bringing about that social revolution towards which history was inevitably proceeding.

These predictions have not been fulfilled: so far from deteriorating progressively the lot of the working classes in the great industrial countries in which their welfare was thought to be most gravely at risk has slowly and progressively improved.

The other great historicist doctrine that has dominated the political thought of the twentieth century is Fascism. The high Fascism of Nazi apologists such as Alfred Rosenberg is a kind of racial or genetic élitism which declares that social progress and the advancement of mankind is rather specially the privilege and responsibility of a racial élite whose inborn superiority is seen to best advantage in the German peoples, who by reason of this inborn superiority will of historic necessity conquer the lesser nations and thus rise to their full stature as rulers of the world.

We still have reason to thank God that this transformation of the world never came about and that the doctrine of genetic élitism is now for the most part confined to morally criminal sects in Britain and the USA. It is not thought likely that it will ever again become a major factor in the causation of wars on a global scale.

It would be a mistake to think that historicism considered as a cultural disease is no longer a threat, for there are several contexts in which historicist thinking still flourishes. The most important of these are in economics and demography, which I shall deal with in turn.

Economic prediction and prediction of the weather

There are certain striking similarities of principle between economic prediction and weather prediction – apart, that is, from the fact that both may be comically wrong. There are important differences too – and these are such as to make prediction of the weather easier to execute and generally speaking more reliable than economic prediction.

The similarities are that both deal with multivariate systems of enormous complexity and of great intrinsic difficulty (a multivariate system is one whose properties are determined by the values of not just one or two but a great number of factors that vary independently and are individually subject to sampling errors and errors of ascertainment). In both the prediction is deductive in style, that is to say the prediction is about the future state of a system of which the present state is known or assumed to be known and the pattern of interaction between the variable quantities is also assumed to be known. The differences are set out in the accompanying table, which I now explain.

Economic and weather prediction compared (see text for explanation of table)

Weather	Economy
Variables scalar	often non-scalar
Functional relations often exactly known	usually not known
Uninfluenced by politics or fashion	influenced by both
Wholly non-reflexive	highly reflexive

1 Weather prediction has to do with scalar variables, that is to say, simply measurable or numerable quantities such as inches of rainfall, wind speeds, per cent cloud cover, humidity, barometric pressure and so on. But many important non-scalar variables enter into economic prediction: confidence, for example. No single factor has a greater influence upon the economy, but how is it to be measured, if at all? Again, economic prediction often depends upon political predictions that are notorious for their fallibility ('In my opinion the Democrats will sweep back into power').

2 In weather prediction the functional relations between variables are physical in character and have been ascertained from long-standing empirical experience – I mean the relations between temperature, barometric pressure, humidity and the likelihood of precipitation. But with the economy comparable functional relations are mostly conjectural in character; their equivalent in the world of meteorology would be propositions such as 'When the wind blows, the cradle will rock.'

3 'Fashion' is almost by definition capricious and unpredictable.

The weather is wholly immune to the vagaries of fashion while the economy is not.

There are special circumstances in which political decisions might influence the weather – for instance, a political decision increasing the number of trials of nuclear weapons. But being themselves unpredictable such considerations do not enter meteorological forecasts; they do, however, enter into the economic forecasts. They are subject to two sources of error: the political presumption may be mistaken and so also may be the nature of the political influence on the economy. Thus the coming into power of a strong right-wing government in Boolooland may strengthen the zlotnik (Bernard Levin's generic term for all units of foreign currency) but the new president's aggressive attitude towards his neighbours may arouse alarm and thus cause the value of the zlotnik to tumble.

4 'Reflexive' is a technical term used in logic and grammar and meaning 'referring to or acting upon itself'. To describe weather prediction as 'non-reflexive' is merely to say, what is self-evidently true, that weather predictions do not affect the weather. It is notorious, though, that economic predictions can affect the economy: economic predictions are strongly reflexive in character. For this reason especially, though there are many others, weather prediction is sounder in principle than economic prediction – is altogether a safer bet. Another circumstance that makes weather prediction easier in principle than economic prediction is the apparent stability of the factors that influence climate, which means that meteorologists can appeal to precedents and historical evidence dating back as far as records exist – for after all (I quote D'Arcy Thompson) 'A snowflake is the same today as when the first snows fell.' It is far otherwise with the economy because the factors that influence the economy change not merely from year to year but even from day to day (think of Middle Eastern oil, for example). In short, the economy has changed far more radically since the days of Adam Smith than the weather has since Benjamin Franklin's.

It amazes me therefore that people who are far too canny and experienced to take weather predictions as certainties or to make their plans and perhaps alter their life-styles in accordance with them are yet almost completely gullible when it comes to economic prediction.

Since they are of great intrinsic difficulty, beset by all the special difficulties that have just been outlined, we cannot wonder that economic predictions are often grievously mistaken.

It is not their wrongness so much as their pretensions to rightness that have brought economic predictions and the theory that underlies them into well-deserved contempt. The dogmatic self-assurance and the asseverative confidence of economists are additional causes of grievance – self-defeating traits among people eager to pass for scientists.

Demographic prediction

Demography is the branch of sociology that has to do with populations: their structure, distribution and reproduction. Short-term demographic predictions are necessary, desirable, and in the main not grossly inaccurate. Knowing the number of births today and assuming that there will be no dramatic change in mortality during the next five or six years, we can predict with adequate accuracy how many places we should provide in primary schools five years hence, or in high schools ten to twelve years hence. This is straightforward enough, but long-term predictions are especially prone to error.

The high tide of population prophecy was in the 1930s when fertility in the countries of northern Europe and the USA had fallen so low as to raise in many minds the question of whether the peoples of the Western world might not die out altogether through non-replacement. Books were written with titles such as *The Twilight of Parenthood* (by Dr Enid Charles), and it is a measure of the gravity of the situation that the world's foremost demographer, the former assistant statistician of the Metropolitan Life Assurance Company, was moved to poetry in commenting a few years later on the situation.

It surely would be a strange trick of fate [he remarked] if we, the most advanced product of organic evolution, should be the first of all living species so clever as to foresee its own doom.

> Like some bold seër in a trance
> Seeing all his own mischance –

We have eaten of the fruit of the Tree of Knowledge, and it has turned to poison.

So wide of the mark were these gloomy prognostications that it was not long before the fear of depopulation was supplanted by our current principal cause of concern: over-population. Human beings are so riotously fertile, it is thought, and so effectively do medicine and hygiene now protect them from infectious disease and other major causes of mortality, that mankind faces a threat of over-population so severe as to put us at the mercy of starvation and new pestilences of unimaginable severity.

The threat of over-population should on no account be under-estimated, but it must be remembered that the predictions which give rise to it do not have the apodictic certainty that was at one time imputed to them, for new and unforeseeable factors have been found to enter into the equations of mortality and fertility, one of them being that an increase in the standard of living may be accompanied by a diminution of fertility rather than by the still greater increase that had at one time been feared. We are by no means out of danger yet but I feel that the period of acute panic is over and has been giving way to a more sober assessment of the practicality and effectiveness of the procedures that might help to keep populations within reasonable bounds.

Learning from past mistakes, as economists sometimes do not, demographic prediction has improved greatly and commands far greater respect now than hitherto. One of the great mistakes made in the past was to suppose that a population's reproductive well-being and the probability of its maintaining its numbers could be measured by a single figure such as the net reproduction ratio or 'Malthusian parameter', much as a patient's degree of fever can be represented by a single thermometer reading – a procedure that has no parallel in demography.[1] Demographic predictions are now based upon the assessment of variables that have a real meaning in terms of the way people actually behave: variables such as marriage rates, marriage ages, ages in life at which children are born – that is, they include the pattern of family building etc. Above all, the *cohort* tends to have supplanted the population as a whole or some sub-category of it as the subject of prediction. A 'cohort' consists, for example, of all the people born in one year or married in one year or going to school in one year, and the fertility and mortality of each cohort can be followed

---

[1] See 'Unnatural Science', pp. 167–83 above.

throughout life until they all drop out. This is sometimes spoken of as a 'longitudinal' as opposed to a cross-sectional way of looking at the population.

Improved though they may be I do not think that we shall ever be able to rely on demographic predictions as we rely upon astronomical predictions, but so much of educational, fiscal, military and political importance turns upon population numbers that we must always try: there can be no abdication from this responsibility, however imperfect the process of prediction may be; but these imperfections are such that we shall probably never be wholly free from fear of over-population or of some degree of depopulation.

Negative predictions

No kind of prediction is more obviously mistaken or more dramatically falsified than that which declares that something which is possible in principle (that is, which does not flout some established scientific law) will never or can never happen. I shall choose now from my own subject, medical science, a bouquet of negative predictions chosen not so much for their absurdity as for the way in which they illustrate something interesting in the history of science or medicine.

My favourite prediction of this kind was made by J. S. Haldane (the distinguished physiologist father of the geneticist J. B. S. Haldane), who in a book published in 1930 titled *The Philosophy of Biology* declared it to be 'inconceivable' that there should exist a chemical compound having exactly the properties since shown to be possessed by deoxyribonucleic acid (DNA). DNA is the giant molecule that encodes the genetic message which passes from one generation to the next – the message that prescribes how development is to proceed. The famous paper in the scientific journal *Nature* in which Francis Crick and James Watson described the structure of DNA and how that structure qualifies it to fulfil its genetic functions was published not so many years after Haldane's unlucky prediction.[1] The possibility that such a compound as DNA might exist had been clearly envisaged by the German nature-philosopher Richard Semon in a book *The Mneme*, a reading of which prompted Haldane to dismiss the whole idea as nonsense.

[1] See 'Lucky Jim', pp. 270–8 above.

In the days before the introduction of antisepsis and asepsis, wound infection was so regular and so grave an accompaniment of surgical operations that we can hardly wonder at the declaration of a well-known surgeon working in London, Sir John Erichsen (1818–96), that 'The abdomen is forever shut from the intrusions of the wise and humane surgeon.' Of course, the coming of aseptic surgery to which I refer below, combined with the improvement of anaesthesia, soon made nonsense of this and opened the door to the great achievements of gastrointestinal surgery in the first decade of our century.

One of the very greatest surgeons of this period was Berkeley George Moynihan of Leeds (1865–1936), a man whose track-record for erroneous predictions puts him in a class entirely by himself.

Around 1900 the famous British periodical, the *Strand* magazine (the first to publish the case records of Sherlock Holmes), thought that at the turn of the century its readers would be interested to know what was in store for them in the century to come; 'a Harley Street surgeon' (unmistakably Moynihan) was accordingly invited to tell them what the future of surgery was to be. Evidently not spectacular, for Moynihan opined that surgery had reached its zenith and that no great advances were to be looked for in the future – nothing as dramatic, for example, as the opening of the abdomen, an event regarded with as much awe as the opening of Japan.

Moynihan's forecast was not the hasty, ill-considered opinion of a busy man: it represented a firmly held conviction. In a Leeds University Medical School magazine in 1930 he wrote: 'We can surely never hope to see the craft of surgery made much more perfect than it is today. We are at the end of a chapter.' Moynihan repeated this almost word for word when he delivered Oxford University's most prestigious lecture, the Romanes Lecture, in 1932. He was a vain and arrogant man, and if these quotations are anything to go by a rather silly one too, but surgery is indebted to him nevertheless, for he introduced the delicacy and fastidiousness of technique that did away for ever with the image of the surgeon as a brusque, over-confident and rough-and-ready sawbones. Morover Moynihan, along with William Stewart Halsted of Johns Hopkins (1852–1922), introduced into modern surgery the *aseptic* technique with all the rituals and drills that go with it: the scrupulous scrub-up, the gown, cap and rubber gloves, and the facial mask over the top of which the pretty young theatre nurse gazes with smouldering eyes at the handsome young intern who is planning to wrong her. These innovations may be said to have made possible the hospital soap opera and thus in turn TV itself – for what would TV be without the hospital drama, and what would the hospital drama be without cap and masks and those long, meaningful stares?

The full regalia of the surgical operation did not escape a certain amount of gentle ridicule – in which we may hear the voice of those older, coarser surgeons whom Moynihan supplanted. Moynihan was once described as 'the pyloric pierrot', and upon seeing Moynihan's rubber shoes a French surgeon is said to have remarked 'Surely he does not intend to stand in the abdomen?'

It might easily be thought that the errors of judgement to which I have called attention are the work of individual men and that committees would be sure to do better. It would be wishful thinking to entertain any such belief. In a report published at the end of the 1914–18 war the leading administrative body of medical high-ups in Great Britain wrote: 'a paraplegic patient may live for a few years in a state of more or less ill-health'.

This remark was quoted a little sadly by the neurologist Ludwig Guttmann in the introduction to his great text on injuries of the spinal cord, *Spinal Cord Injuries: Comprehensive Management and Research* – a work written while he was planning the first paraplegic Olympic Games in Israel.

In the appraisal of all these misjudgements we must not forget that it is the erroneous prediction that tends to stick in the mind: there is not the same incentive to remember the predictions that are right, because if they are self-evidently true the likelihood is that they are also rather dull. Thus the great Canadian physician Sir William Osler always insisted that chemistry would play an increasingly important part in modern medicine – and he could hardly have been more right.

Nevertheless, some medical high-ups do seem to be rather especially prone to making confident predictions. I believe this to be due to the steeply hierarchical nature of the medical profession, which makes the high-ups so accustomed to the deference and even sycophancy of medical students, nurses and their juniors generally as to engender in them the idea that their opinions are especially likely to be right because it is they who hold them.

The philosophy of prediction

In the discussion of astronomical predictions – among which I include all predictions that turn on logical deductions from premises believed, hoped or assumed to be true – there is no element of fore-knowledge at all: the prediction is based upon *present* knowledge which of course may or may not be right.

I now propose to show that the most interesting – and what would be the most valuable – kind of prediction, the prediction of future ideas, is not logically possible, and that it is a fallacy to suppose that although long-term predictions are highly fallible, short-term predictions can be exactly right.

The arguments I shall bring forward turn upon simple logical paradoxes, the evidence of which is conclusive, for any line of reasoning that embodies a self-contradiction must certainly be mistaken. On no account belittle the paradox, therefore, or dismiss it as a mere play upon words: a paradox has the same significance for the logician as the smell of burning rubber has for the electronics engineer.

First consider the self-contradictoriness of supposing that one can predict future ideas ('future' means a day, a week, a year or even ten minutes hence). Consider the following statement: 'I predict that in 1983 research workers at New York's Memorial Sloan-Kettering Cancer Center will propound the entirely novel theory of the role of viruses in the causation of cancer, to wit . . .'. There are now only two possibilities. The statement can be completed or it can be left incomplete. If the statement is completed the theory whose promulgation I predict cannot be wholly new in 1983 because to complete the sentence I shall have to propound it here and now, so it would not be a future idea at all; if on the other hand I leave the declaration incomplete then I am not making a prediction. (It was considerations such as these that led Karl Popper to refute the pretensions of the cultural disease of modern sociology that he described as *historicism*.)

It might be maintained that whereas long-term predictions are altogether too hazardous for us to have any confidence in them, nevertheless short-term predictions can sometimes be dead accurate. But this leads to a paradox too, for accurate short-term prediction implies accurate long-term prediction, which we have just conceded to be impossible. Consider for example a prediction made in 1982 on any topic whatsoever relating to the year 1990, and let us agree that it is impossible that so distant a prediction should be accurate. If, however, we could predict accurately all that will happen as soon as next year, this prediction will include an accurate statement of what someone in 1982 is going to say about what will happen in 1983; this in turn will include what someone in 1983 is going to say about what will happen in 1984, and so on until we reach 1990, and we shall end by having made the accurate long-term prediction that we have declared to be impossible. Accurate short-term prediction logically entails accurate long-term prediction – a palpable self-contradiction.

Advice to a young reader

My own interest in prediction grew out of the series of articles in the *Strand* magazine to which I referred above, a series in which various eminent people were invited at the turn of the century to forecast what would happen in the following century.

The turn of the millennium (2000/2001) will make severe demands upon soothsayers and I expect some readers of this article will be invited to make pronouncements upon, for example, 'air travel/psychiatry/podiatry in the third millennium AD'. I should like to give literally one word of advice to all who are invited to make such predictions: *Don't.*

SCIENCE AND THE
SANCTITY OF LIFE

I DO not intend to deny that the advance of science may sometimes have consequences that endanger, if not life itself, then the quality of life or our self-respect as human beings (for it is in this wider sense that I think 'sanctity' should be construed). Nor shall I waste time by defending science as a whole or scientists generally against a charge of inner or essential malevolence. The Wicked Scientist is not to be taken seriously: Dr Strangelove, Dr Moreau, Dr Moriarty, Dr Mabuse, Dr Frankenstein (an honorary degree, this) and the rest of them are puppets of Gothic fiction. Scientists, on the whole, are amiable and well-meaning creatures. There must be very few wicked scientists. There are, however, plenty of wicked philosophers, wicked priests and wicked politicians.

One of the gravest charges ever made against science is that biology has now put it into our power to corrupt both the body and the mind of man. By scientific means (the charge runs) we can now breed different kinds and different races – different 'makes', almost – of human beings, degrading some, making aristocrats of others, adapting others still to special purposes: treating them in fact like dogs, for this is how we *have* treated dogs. Or again: science now makes it possible to dominate and control the thought of human beings – to improve them, perhaps, if that should be our purpose, but more often to enslave or to corrupt with evil teaching.

But these things have always been possible. At any time in the past five thousand years it would have been within our power to embark on a programme of selecting and culling human beings and raising breeds as different from one another as toy poodles and Pekinese are from St Bernards and Great Danes. In a genetic sense the empirical arts of the breeder are as easily applicable to human beings as to horses – more easily applicable, in fact, for human beings are highly *evolvable* animals, a property they owe partly to an open and uncomplicated breeding system, which allows them a glorious range of inborn diversity and therefore a tremendous evolutionary potential; and partly to their lack of physical specialisations (in the sense in which ant-eaters and wood-peckers and indeed dogs are specialised), a property which gives human beings a sort of amateur status among animals. And it has always been possible to pervert or corrupt human beings by coercion, propaganda or evil indoctrination. Science has not yet improved these methods, nor have scientists used them. They have, however, been used to great effect by politicians, philosophers and priests.

The mischief that science may do grows just as often out of trying to do good – as, for example, improving the yield of soil is intended to do good – as out of actions intended to be destructive. The reason is simple enough: however hard we try, we do not and sometimes cannot foresee all the distant consequences of scientific innovation. No one clearly foresaw that the widespread use of antibiotics might bring about an evolution of organisms resistant to their action. No one could have predicted that X-irradiation was a possible cause of cancer. No one could have foreseen the speed and scale with which advances in medicine and public health would create a problem of over-population that threatens to undo much of what medical science has worked for. (Thirty years ago the talk was all of how the people of the Western world were reproducing themselves too slowly to make good the wastage of mortality; we heard tell of a 'Twilight of Parenthood', and wondered rather fearfully where it all would end.) But somehow or other we shall get round all these problems, for every one of them is soluble, even the population problem, and even though its solution is obstructed above all else by the bigotry of some of our fellow men.

– I choose from medicine and medical biology one or two concrete examples of how advances in science threaten or seem to threaten the sanctity of human life. Many of these threats, of course, are in no sense distinctively medical, though they are often loosely classified as such. They are merely medical contexts for far more pervasive dangers. One of them is our increasing state of dependence on medical services and the medical industries. What would become of the diabetic if the supplies of insulin dried up, or of the victims of Addison's disease deprived of synthetic steroids? Questions of this kind might be asked of every service that society provides. In a complex society we all sustain and depend upon each other – for transport, communications, food, goods, shelter, protection and a hundred other things. The medical industries will not break down all by themselves, and if they do break down it will be only one episode of a far greater disaster.

The same goes for the economic burden imposed by illness in any community that takes some collective responsibility for the health of its citizens. All shared burdens have a cost which is to a greater or lesser degree shared between us: education, pensions, social welfare, legal aid and every other social service, including government.

We are getting nearer what is distinctively medical when we ask ourselves about the economics, logistics and morality of keeping people alive by medical intervention and medical devices. At present it is the cost and complexity of the operation, and the shortage of machines and organs, that denies a kidney graft or an artificial kidney to anyone mortally in need of it. The limiting factors are thus still economic and logistic. But what about the morality of keeping people alive by these heroic medical contrivances? I do not think it is possible to give any answer that is universally valid or that, if it were valid, would remain so for more than a very few years. Medical contrivances extend all the way from pills and plasters and bottles of tonic to complex mechanical prostheses, which will one day include mechanical hearts. At what point shall we say we are wantonly interfering with nature and prolonging life beyond what is proper and humane?

In practice the answer we give is founded not upon abstract moralising but upon a certain natural sense of the fitness of things, a feeling that is shared by most kind and reasonable people even if we cannot define it in philosophically defensible or legally accountable terms. It is only at international conferences that we tend to adopt the convention that people behave like idiots unless acting upon clear and well-turned instructions to behave sensibly. There is in fact no general formula or smooth form of words we can appeal to when in perplexity.

Moreover, our sense of what is fit and proper is not something fixed, as if it were inborn and instinctual. It changes as our experience grows, as our understanding deepens, and as we enlarge our grasp of possibilities – just as living religions and laws change, and social structures and family relationships.

I feel that our sense of what is right and just is already beginning to be offended by the idea of taking great exertions to keep alive grossly deformed or monstrous newborn children, particularly if their deformities of body or mind arise from major defects of the genetic apparatus. There are in fact scientific reasons for changing an opinion that might have seemed just and reasonable a hundred years ago.

Everybody takes it for granted, because it is so obviously true, that a married couple will have children of very different kinds and constitutions on different occasions. But the traditional opinion, which most of us are still unconsciously guided by, is that the child conceived on any one occasion is the unique and necessary product of that occasion: *that* child would have been conceived, we tend to think, or no child at all. This interpretation is quite false, but human dignity and security clamour for it. A child sometimes wonderingly acknowledges that he would never have been born at all if his mother and father had not chanced to meet and fall in love and marry. He does not realise that, instead of conceiving him, his parents might have conceived any one of a hundred thousand other children, all unlike each other and unlike himself. Only over the past one hundred years has it come to be realised that the child conceived on any one occasion belongs to a vast cohort of Possible Children, any one of whom might have been conceived and born if a different spermatozoon had chanced to fertilise the mother's egg cell – and the egg cell itself is only one of very many. It is a matter of luck then, a sort of genetic lottery. And sometimes it is cruelly bad luck – some terrible genetic conjunction, perhaps, which once in ten or twenty thousand times will bring together a matching pair of damaging recessive genes. Such a misfortune, being the outcome of a random process, is, considered in isolation, completely and essentially pointless. It is not even strictly true to say that a particular inborn abnormality must have lain within the genetic potentiality of the parents, for the malignant gene may have arisen *de novo* by mutation. The whole process is unhallowed – is, in the older sense of that word, profane.[1]

---

[1] An eminent theologian once said to me that I was making altogether too much fuss about this kind of mischance. It was, he said, all in the nature of things and already comprehended within our way of thinking; it was not different in principle from being accidentally struck on the head by a falling roof tile. But I think there *is* an important difference of principle. In the process by which a chromosome is allotted to one germ cell rather than another, and in the union of

I am saying that if we feel ourselves under a moral obligation to make every possible exertion to keep a monstrous embryo or new born child alive *because* it is in some sense the naturally intended – and therefore the unique and privileged – product of its parents' union at the moment of its conception, then we are making an elementary and cruel blunder: for it is *luck* that determines which one child is in fact conceived out of the cohort of Possible Children that might have been conceived by those two parents on that occasion. I am not using the word 'luck' of conception as such, nor of the processes of embryonic and foetal growth, nor indeed in any sense that derogates from the wonder and awe in which we hold processes of great complexity and natural beauty which we do not fully understand; I am simply using it in its proper sense and proper place.[1]

This train of thought leads me directly to eugenics – 'the science', to quote its founder, Francis Galton, 'which deals with all the in-fluences that improve the inborn qualities of a race; also with those that develop them to the utmost advantage'. Because the upper and lower boundaries of an individual's capability and performance are set by his genetic make-up, it is clear that if eugenic policies were to be ill-founded or mistakenly applied they could offer a most terrible threat to the sanctity and dignity of human life. This threat I shall now examine.

Eugenics is traditionally subdivided into positive and negative eugenics. Positive eugenics has to do with attempts to improve human beings by genetic policies, particularly policies founded upon selective or directed breeding. Negative eugenics has the lesser ambition of attempting to eradicate as many as possible of our inborn imperfections. The distinction is useful and pragmatically valid for the following reasons.[2] Defects of the genetic constitution (such as those which manifest themselves as Down's syndrome ('mongolism'), haemophilia,

---

germ cells, luck is of the very essence. The random element is an integral, indeed a defining, characteristic of Mendelian inheritance. All I am saying is that it is difficult to wear a pious expression when the fall of the dice produces a child that is structurally or biochemically crippled from birth or conception.

[1] There are, perhaps, weighty legal and social reasons why even tragically deformed children should be kept alive (for who is to decide? and where do we draw the line?), but these are outside my terms of reference.

[2] See my book *The Future of Man* (New York, 1959; London, 1960).

galactosaemia, phenylketonuria and a hundred other hereditary abnormalities) have a much simpler genetic basis than desirable characteristics like beauty, high physical performance, intelligence or fertility.
This is almost self-evident. All geneticists believe that 'fitness' in its
most general sense depends on a nicely balanced co-ordination and
interaction of genetic factors, itself the product of laborious and long
drawn out evolutionary adjustment. It is inconceivable, indeed self-
contradictory, that an animal should evolve into the possession of some
complex pattern of interaction between genes that made it inefficient,
undesirable, or unfit – that is, *less* well adapted to the prevailing circumstances. Likewise, a motor car will run badly for any one of a multitude
of particular and special reasons, but runs well because of the har-
monious mechanical interactions made possible by a sound and
economically viable design.

Negative eugenics is a more manageable and understandable enterprise than positive eugenics. Nevertheless, many well-meaning people
believe that, with the knowledge and skills already available to us, and
within the framework of a society that upholds the rights of individuals,
it is possible in principle to raise a superior kind of human being by a
controlled or 'recommended' scheme of mating and by regulating the
number of children each couple should be allowed or encouraged to
have. If stockbreeders can do it, the argument runs, why should not
we? – for who can deny that domesticated animals have been improved
by deliberate human intervention?

I think this argument is unsound for a lesser and for a more important reason.

1 Domesticated animals have not been 'improved' in the sense in
which we should use that word of human beings. They have not enjoyed an all-round improvement, for some special characteristics or
faculties have been so far as possible 'fixed' without special regard to and
sometimes at the expense of others. Tameness and docility are most
easily achieved at the expense of intelligence, but that does not matter
if what we are interested in is, say, the quality and yield of wool.

2 The ambition of the stockbreeder in the past, though he did not
realise it, was twofold: not merely to achieve a predictably uniform
product by artificial selection, but also to establish an internal genetic

uniformity (homozygosity) in respect of the characters under selection, to make sure that the stock would 'breed true' – for it would be a disaster if characters selected over many generations were to be irrecoverably mixed up in a hybrid progeny. The older stockbreeder believed that uniformity and breeding true were characteristics that necessarily went together, whereas we now know that they can be separately achieved. And he expected his product to fulfil two quite distinct functions which we now know to be separable, and often better separated: on the one hand, to be in themselves the favoured stock and the top performers – the super-sheep or super-mice – and, on the other hand, to be the parents of the next generation of that stock. It is rather as if Rolls-Royces, in addition to being an end-product of manufacture, had to be so designed as to give rise to Rolls-Royce progeny.[1]

It is just as well these older views are mistaken, for with naturally outbreeding populations such as our own, genetic uniformity, arrived at and maintained by selective inbreeding, is a highly artificial state of affairs with many inherent and ineradicable disadvantages.

Stockbreeders, under genetic guidance, are now therefore inclining more and more towards a policy of deliberate and nicely calculated cross-breeding. In the simplest case, two partially inbred and internally uniform stocks are raised and perpetuated to provide two uniform lineages of parents, but the eugenic goal, the marketable end-product or high performer, is the progeny of a cross between members of the two parental stocks. Being of hybrid make-up, the progeny do not breed true, and are not in fact bred from; they can be likened to a manufactured end-product: but they can be uniformly reproduced at will by crossing the two parental stocks. Many more sophisticated regimens of cross-breeding have been adopted or attempted, but the innovation of principle is the same. (1) The end-products are all like each other and are faithfully reproducible, but are not bred from because they do not breed true: the organisms that represent the eugenic goal have been relieved of the responsibility of reproducing themselves. And (2) the end-products, though uniform in the sense of

---

[1] These arguments are set out more fully in my 'The Genetic Improvement of Man' in *The Hope of Progress* (London, 1972), pp. 69–76.

being like each other, are to a large extent hybrid – heterozygous as opposed to homozygous – in genetic composition.

The practices of stockbreeders can therefore no longer be used to support the argument that a policy of positive eugenics is applicable in principle to human beings in a society respecting the rights of individuals. The genetical manufacture of supermen by a policy of cross-breeding between two or more parental stocks is unacceptable today, and the idea that it might one day become acceptable is unacceptable also.

A deep fallacy does in fact eat into the theoretical foundations of positive eugenics and that older conception of stockbreeding out of which it grew.[1] The fallacy was to suppose that the *product* of evolution, that is the outcome of an episode of evolutionary change, was a new and improved genetic formula (genotype) which conferred a higher degree of adaptedness on the individuals that possessed it. This improved formula, representing a new and more successful solution of the problems of remaining alive in a hostile environment, was thought to be shared by nearly all members of the newly evolved population, and to be stable except in so far as further evolution might cause it to change again. Moreover, the population would have to be predominantly homozygous in respect of the genetic factors entering into the new formula, for otherwise the individuals possessing it would not breed true to type, and everything natural selection had won would be squandered in succeeding generations.

Most geneticists think this view mistaken. It is *populations* that evolve, not the lineages and pedigrees of old-fashioned evolutionary 'family trees', and the end-product of an evolutionary episode is not a new genetic formula enjoyed by a group of similar individuals, but a new spectrum of genotypes, a new pattern of genetic inequality, definable only in terms of the population as a whole. Naturally outbreeding populations are not genetically uniform, even to a first approximation. They are persistently and obstinately diverse in respect of nearly all constitutive characters which have been studied deeply enough to say for certain whether they are uniform or not. It is the *population* that breeds true, not its individual members. The progeny of a given population are themselves a population with the same pattern of genetic

[1] See 'A Biological Retrospect', pp. 287–97 above.

make-up as their parents – except in so far as evolutionary or selective forces may have altered it. Nor should we think of uniformity as a desirable state of affairs which *we* can achieve even if nature, unaided, cannot. It is inherently undesirable, for a great many reasons.

The goal of positive eugenics, in its older form, cannot be achieved, and I feel that eugenic policy must be confined (paraphrasing Karl Popper) to *piecemeal genetic engineering*. That is just what negative eugenics amounts to; and now, rather than to deal in generalities, I should like to consider a concrete eugenic problem and discuss the morality of one of its possible solutions.

Some 'inborn' defects – some defects that are the direct consequence of an individual's genetic make-up as it was fixed at the moment of conception – are said to be of *recessive* determination. By a recessive defect is meant one that is caused by, to put it crudely, a 'bad' gene that must be present in *both* the gametes that unite to form a fertilised egg, that is, in both spermatozoon and egg cell, not just in one or the other. If the bad gene *is* present in only one of the gametes, the individual that grows out of its fusion with the other is said to be a *carrier* (technically, a heterozygote).

Recessive defects are individually rather rare – their frequency is of the order of 10^{-4} (one in ten thousand) – but collectively they are most important. Among them are, for example, phenylketonuria, a congenital inability to handle a certain dietary constituent, the amino acid phenylalanine, a constituent of many proteins; galactosemia, another inborn biochemical deficiency, the victims of which cannot cope metabolically with galactose, an immediate derivative of milk sugar; and, more common than either, fibrocystic disease of the pancreas, believed to be the symptom of a generalised disorder of mucus-secreting cells. All three are caused by particular genetic defects; but their secondary consequences are manifold and deep-seated. The phenylketonuric baby is on the way to becoming an imbecile. The victim of galactosemia may become blind through cataract and be mentally retarded.

Contrary to popular superstition, many congenital ailments can be prevented or, if not prevented, cured. But in this context prevention and cure have very special meanings.

The phenylketonuric or galactosemic child may be protected from the consequences of his genetic lesion by keeping him on a diet free from phenylalanine in the one case or lactose in the other. This is a most unnatural proceeding, and much easier said than done, but I take it no one would be prepared to argue that it was an unwarrantable interference with the workings of providence. It is not a cure in the usual medical sense because it neither removes nor repairs the underlying congenital deficiency. What it does is to create around the patient a special little world, a microcosm free from phenylalanine or galactose as the case may be, in which the genetic deficiency cannot express itself outwardly.

Now consider the underlying morality of prevention. We can prevent phenylketonuria by preventing the genetic conjunction responsible for it in the first instance, that is, by preventing the coming together of an egg cell and a sperm each carrying that same one harmful recessive gene. All but a very small proportion of overt phenylketonurics are the children of parents who are both carriers – carriers, you remember, being the people who inherited the gene from one only of the two gametes that fused at their conception. Carriers greatly outnumber the overtly afflicted. When two carriers of the same gene marry and bear children, one-quarter of their children (on the average) will be normal, one-quarter will be afflicted, and one-half will be carriers like themselves. We shall accomplish our purpose, therefore, if, having identified the carriers (another thing easier said than done, but it *can* be done, and in an increasing number of recessive disorders), we try to discourage them *from marrying each other* by pointing out the likely consequences if they do so. The arithmetic of this is not very alarming. In a typical recessive disease, about one marriage in every five or ten thousand would be discouraged or warned against, and each disappointed party would have between fifty and a hundred other mates to choose from.

If this policy were to be carried out, the overt incidence of a disease like phenylketonuria, in which carriers can be identified, would fall almost to zero between one generation and the next.

Nevertheless the first reaction to such a proposal may be one of outrage. Here is medical officiousness planning yet another insult to human dignity, yet another deprivation of the rights of man. First it was vaccination and then fluoride; if now people are not to be allowed to marry whom they please, why not make a clean job of it and over-throw the Crown or the United States Constitution?

But reflect for a moment. What is being suggested is that a certain small proportion of marriages should be discouraged for genetic reasons, to help us do our best to avoid bringing into the world children who are biochemically crippled. In all cultures marriages are already pro-hibited for genetic reasons – the prohibition, for example, of certain degrees of inbreeding (the exact degree varies from one culture or religion to another). Thus the prohibition of marriage has an im-memorial authority behind it. As to the violation of human dignity entailed by performing tests on engaged couples that are no more complex or offensive than blood tests, let me say only this: if anyone thinks or has ever thought that religion, wealth or colour are matters that may properly be taken into account when deciding whether or not a certain marriage is a suitable one, then let him not dare to suggest that the genetic welfare of human beings should not be given equal weight.

I think that engaged couples should themselves decide, and I am pretty certain they would be guided by the thought of the welfare of their future children. When it came to be learned, about twenty years ago, that marriages between Rhesus-positive men and Rhesus-negative women might lead to the birth of children afflicted by haemolytic disease, a number of young couples are said to have ended their engagements – needlessly, in most cases, because the dangers were overestimated through not being understood. But that is evidence enough that young people marrying today are not likely to take a stand upon some hypothetical right to give birth to defective children, if, by taking thought, they can do otherwise.

The problems I have been discussing illustrate very clearly the way in which scientific evidence bears upon decisions that are not, of course, in themselves scientific. If the termination of a pregnancy is now in question, scientific evidence may tell us that the chances of a defective birth are 100 per cent, 50 per cent, 25 per cent, or perhaps unascertain-able. The evidence is highly relevant to the decision, but the decision itself is not a scientific one, and I see no reason why scientists as such should be specially well qualified to make it. The contribution of science is to have enlarged beyond all former bounds the evidence we must take account of before forming our opinions. Today's opinions may not be the same as yesterday's, because they are based on fuller or better evidence. We should quite often have occasion to say 'I used to think that once, but now I have come to hold a rather different opinion.' People who never say as much are either ineffectual or dangerous.

We all nowadays give too much thought to the material blessings or evils that science has brought with it, and too little to its power to liberate us from the confinements of ignorance and superstition.

It may be that the greatest liberation of thought ever achieved by the scientific revolution was to have given mankind the expectation of a future in this world. The idea that the world has a virtually indeterminate future is comparatively new. Much of the philosophic speculation of three hundred years ago was oppressed by the thought that the world had run its course and was coming shortly to an end.[1] 'I was borne in the last Age of the World,' said John Donne, giving it as the 'ordinarily received' opinion that the world had thrice two thousand years to run between its creation and the Second Coming. According to Archbishop Ussher's chronology more than five and a half of those six thousand years had gone by already.[2]

No empirical evidence challenged this dark opinion. There were no new worlds to conquer, for the world was known to be spherical and therefore finite; certainly it was not all known, but the full extent of what was *not* known was known. Outer space did not put into people's minds then, as it does into ours now, the idea of a tremendous endeavour just beginning.

Moreover, life itself seemed changeless. The world a man saw about him in adult life was much the same as it had been in his own childhood, and he had no reason to think it would change in his own or his children's lifetime. We need not wonder that the promise of the next world was held out to believers as an inducement to put up with the incompleteness and inner pointlessness of this one: the present world was only a staging post on the way to better things. There was a certain

[1] See Stephen Toulmin and June Goodfield, *The Discovery of Time* (London and New York, 1965).
[2] 'We are almost the last progeny of the First Men,' said Thomas Burnet.

awful topicality about Thomas Burnet's description of the world in flames at the end of its long journey from 'a dark chaos to a bright star', for the end of the world might indeed come at any time. And Thomas Browne warned us against the folly and extravagance of raising monuments and tombs intended to last for many centuries. We are living in The Setting Part of Time, he told us: *the Great Mutations of the World are acted: it is too late to be ambitious.*

Science has now made it the ordinarily received opinion that the world has a future reaching beyond the most distant frontiers of the imagination – and that is perhaps why, in spite of all his faults, so many scientists still count Francis Bacon their first and greatest spokesman: we may yet build a New Atlantis. The point is that when Thomas Burnet exhorted us to become 'Adventurers for Another World' *he* meant the next world – but we mean this one.

ON 'THE EFFECTING OF
ALL THINGS POSSIBLE'

1

My title, or, if you like, my motto, comes from Francis Bacon's *New Atlantis*, published in 1627. The *New Atlantis* was Bacon's dream of what the world might have been, and might still become, if human knowledge were directed towards improving the worldly condition of man. It makes a rather strange impression nowadays, and very few people bother with it who are not interested either in Bacon himself, or in the flux of seventeenth-century opinion or the ideology of Utopias. We shall not read it for its sociological insights, which are non-existent, nor as science fiction, because it has a general air of implausibility; but there is one high poetic fancy in the *New Atlantis* that stays in the mind after all its fancies and inventions have been forgotten. In the New Atlantis, an island kingdom lying in very distant seas, the only commodity of external trade is – *light*: Bacon's own special light, the light of understanding. The Merchants of Light who carry out its business are members of a society or order of philosophers who between them make up (so their spokesman declares) 'the noblest foundation that ever was upon the earth'. 'The end of our foundation', the spokesman goes on to say, 'is the knowledge of causes and the secret motions of things; and the enlarging of the bounds of human empire, to the effecting of all things possible.' You will see later on why I chose this motto.

2

My purpose is to draw certain parallels between the spiritual or philosophic condition of thoughtful people in the seventeenth century and in the contemporary world, and to ask why the great philosophic revival that brought comfort and a new kind of understanding to our predecessors has now apparently lost its power to reassure us and cheer us up.

The period of English history that lies roughly between the accession of James I in 1603 and the English Civil War has much in common with the present day.[1] For the historian of ideas, it is a period of questioning and irresolution and despondency; of sermonising but also of satire; of rival religions competing for allegiance, among them the 'black doctrine of absolute reprobation'; a period during which our human propensity towards hopefulness was clouded over by a sense of inconstancy and decay. Literary historians have spoken of a 'metaphysical shudder',[2] and others of a sense of crisis or of a 'failure of nerve'.[3] Of course, we must not imagine that ordinary people went around with the long sunk-in faces to be expected in the victims of a spiritual deficiency disease. It was philosophic or reflective man who had these misgivings, the man who is all of us some of the time but none of us all of the time, and we may take it that, then as now, the remedy for discomforting thoughts was less often to seek comfort than to abstain from thinking.

Amidst the philosophic gloom of the period I am concerned with, new voices began to be heard which spoke of hope and of the possibility of a future (a subject I shall refer to later on); which spoke of confidence in human reason, and of what human beings might achieve through an understanding of nature and a mastery of the physical world. I think there can be no question that, in this country, it was Francis Bacon who started the dawn chorus – the man who first defined the newer purposes of learning and, less successfully, the means by which they might be fulfilled. Human spirits began to rise. To use a good old seventeenth-century metaphor, there was a slow change, but ultimately a complete one, in the 'climate of opinion'. It became no longer the thing to mope. In a curious way the Pillars of Hercules – the 'Fatal Columns' guarding

[1] See, for example, Herbert Grierson, *Cross Currents in English Literature of the Seventeenth Century* (London, 1929); Basil Willey, *The Seventeenth-Century Background* (London, 1934); G. N. Clark, *The Seventeenth Century*, 2nd ed. (Oxford, 1947); W. Notestein, *The English People on the Eve of Colonization* (New York, 1954); Marjorie Hope Nicolson, *Mountain Gloom and Mountain Glory* (Ithaca, 1959); Maurice Ashley, *England in the Seventeenth Century*, 3rd ed. (London, 1961); H. R. Trevor-Roper, *Religion, the Reformation and Social Change* (London, 1967).

[2] George Williamson, 'Mutability, Decay and Seventeenth Century Melancholy', *Journal of English Literature and History*, **2**, 121–50, 1935.

[3] Christopher Hill, *Intellectual Origins of the English Revolution* (Oxford, 1965).

the Straits of Gibraltar that make the frontispiece to Bacon's *Great Instauration* – provided the rallying cry of the New Philosophy. Let me quote a great American scholar's, Dr Marjorie Hope Nicolson's,[1] description of how this came about:

Before Columbus set sail across the Atlantic, the coat of arms of the Royal Family of Spain had been an *impressa*, depicting the Pillars of Hercules, the Straits of Gibraltar, with the motto, *Ne Plus Ultra*. There was 'no more beyond'. It was the glory of Spain that it was the outpost of the world. When Columbus made his discovery, Spanish Royalty thriftily did the only thing necessary: erased the negative, leaving the Pillars of Hercules now bearing the motto, *Plus Ultra*. There was more beyond . . .

And so *plus ultra* became the motto of the New Baconians, and the frontispiece to the *Great Instauration* shows the Pillars of Hercules with ships passing freely to and fro.

One symptom of the new spirit of enquiry was, of course, the foundation of the Royal Society and of sister academies in Italy and France. That story has often been told, and in more than one version, because the parentage of the Royal Society is still in question.[2] We shall be taking altogether too narrow a view of things, however, if we suppose that the great philosophic uncertainties of the seventeenth century were cleared up by the fulfilment of Bacon's ambitions for science. Modern scientific research began earlier than the seventeenth century.[3] The great achievement of the latter half of the seventeenth century was to arrive at a general scheme of belief within which the cultivation of science was seen to be very proper, very useful, and by no means irreligious. This larger conception or purpose, of which science was a principal agency, may be called 'rational humanism' if we are temperamentally in its favour and take our lead from the writings of John Locke, or 'materialistic rationalism' if we are against it and frown disapprovingly over Thomas Hobbes, but neither descrip-

[1] Marjorie Hope Nicolson, 'Two Voices: Science and Literature', *Rockefeller Review*, 3, 1–11, 1963.

[2] See, for example, Margery Purver, *The Royal Society: Concept and Creation* (London, 1967) and a number of papers in *Notes and Records of the Royal Society*, 23, no. 2, December 1968.

[3] For England in particular, see Christopher Hill, op. cit. (p. 325 above, note 3); F. R. Johnson, *Astronomical Thought in Renaissance Britain* (Baltimore, 1937).

tion is satisfactory, because the new movement had not yet taken on the explicit character of an alternative or even an antidote to religion, which is the sense that 'rational humanism' tends to carry with it today.

However we may describe it, rational humanism became the dominant philosophic influence in human affairs for the next 150 years, and by the end of the eighteenth century the spokesmen of Reason and Enlightenment – men such as Adam Ferguson and William Godwin and Condorcet – take completely for granted many of the ideas that had seemed exhilarating and revolutionary in the century before. But over this period an important transformation was taking place. The seventeenth-century doctrine of the *necessity* of reason was slowly giving way to a belief in the *sufficiency* of reason – so illustrating the tendency of many powerful human beliefs to develop into an extreme or radical form before they lose their power to persuade us, and in doing so to create anew many of the evils for which at one time they professed to be the remedy. (It has often been said that rationalism in its more extreme manifestations could only supplant religion by acquiring some of the characteristics of religious belief itself.) Please don't interpret these remarks as any kind of attempt to depreciate the power of reason. I emphasise the distinction between the ideas of the necessity and of the sufficiency of reason as a defence against that mad and self-destructive form of anti-rationalism which seems to declare that because reason is not sufficient, it is not necessary.

Many reflective people nowadays believe that we are back in the kind of intellectual and spirtual turmoil that disturbed the first half of the seventeenth century. Both epochs are marked, not by any characteristic system of beliefs (neither can be called 'The Age of' anything), but by an equally characteristic syndrome of unfixed beliefs; by the emptiness that is left when older doctrines have been found wanting and none has yet been found to take their place. Both epochs have the characteristics of a philosophic interregnum. In the first half of the seventeenth century, the essentially medieval world-picture of Elizabethan England had lost its power to satisfy and bring comfort, just as nowadays the radical materialism traditionally associated with Victorian thinkers seems quite inadequate to remedy our complaints. By a curious inversion of thinking, scholastic reasoning is said to have failed because it discouraged new enquiry, but that was precisely the measure of its success. For that is just what successful, satisfying explanations do: they confer a sense of finality; they remove the incentive to work things out anew. At all events the repudiation of Aristotle and the hegemony of ancient learning, of the scholastic style of reasoning, of the illusion of a Golden Age, is as commonplace in the writings of the seventeenth century as dismissive references to rationalism and materialism in the literature of the past fifty years.

We can draw quite a number of detailed correspondences between the contemporary world and the first forty or fifty years of the seventeenth century, all of them part of a syndrome of dissatisfaction and unbelief; and though we might find reason to cavil at each one of them individually, they add up to an impressive case. Novels and philosophical *belles-lettres* have now an inward-looking character, a deep concern with matters of personal salvation and a struggle to establish the authenticity of personal existence; and we may point to the prevalence of satire and of the Jacobean style of 'realism' on the stage. I shall leave aside the political and economic correspondences between the two epochs,[1] important though they are, and confine myself to analogies that might be described as 'philosophical' in the homely older sense, the sense that has to do with the purpose and conduct of life and with the attempt to answer the simple questions that children ask. Once again we are oppressed by a sense of decay and deterioration, but this time, in part at least, by a fear of the deterioration of the world through technological innovation. Artificial fertilisers and pesticides are undermining our health (we tell ourselves), soil and sea are being poisoned by chemical and radioactive wastes, drugs substitute one kind of disease for another, and modern man is under the influence of stimulants whenever he is not under the influence of sedatives. Once again there is a feeling of despondency and incompleteness, a sense of doubt about the adequacy of man, amounting in all to what a future historian might again describe as a failure of nerve. Intelligent and

---

[1] England at the time of the Armada was a prosperous country, and it became so again in the reign of Queen Anne; the period I am discussing, however, was marked by a high level of unemployment and a number of major economic slumps, not to mention the English Civil War; moreover the reputation of England abroad sank to a specially low level in the latter part of James I's reign and during the reign of Charles I. This was also the period of the great emigrations to Massachusetts.

learned men may again seek comfort in an elevated kind of barminess (but something kind and gentle nevertheless). Mystical syntheses between science and religion, like the Cambridge Neoplatonism of the mid-seventeenth century, have their counterpart today, perhaps, in the writings and cult of Teilhard de Chardin and in a revival of faith in the Wisdom of the East. Once again there is a rootlessness or ambivalence about philosophical thinking, as if the discovery or rediscovery of the insufficiency of reason had given a paradoxical validity to nonsense, and this gives us a special sympathy for the dilemmas of the seventeenth century. To William Lecky, the great nineteenth-century historian of rationalism, it seemed almost beyond comprehension that witch hunting and witch burning should have persisted far into the seventeenth century, or that Joseph Glanvill should have been equally an advocate of the Royal Society and of belief in witchcraft.[1]

We do not wonder at it now. It no longer seems strange to us that Pascal the geometer who spoke with perfect composure about infinity and the infinitesimal should have been supplanted by Pascal the great cosmophobe who spoke with anguish about the darkness and loneliness of outer space. Discoveries in astronomy and cosmology have always a specially disturbing quality. We remember the dismay of John Donne and Pascal himself and latterly of William Blake. Cosmological discoveries bring with them a feeling of awe but also, for most people, a sense of human diminishment. Our great sidereal adventures today are both elevating and frightening, and may be both at the same time. The launching of a space rocket is (to go back to seventeenth-century language) a tremendous phenomenon. It must have occurred to many who saw pictures of it that the great steel rampart or nave from which the Apollo rockets were launched had the size and shape and grandeur of a cathedral, with Apollo itself in the position of a spire. Like a cathedral it is economically pointless, a shocking waste of public money; but like a cathedral it is also a symbol of aspiration towards higher things.

When we compare the climates of opinion in the seventeenth century and today, we must again remember that cries of despair are not necessarily authentic. There was a strong element of affectation

[1] William Lecky, *The Rise and Influence of Rationalism in Europe* (London, 1865); see especially H. R. Trevor-Roper, op. cit. (p. 325 above, note 1).

about Jacobean melancholy, and so there is today. Then as now it had tended to become a posture. One of a modern writer's claims to be taken seriously is to castigate complacency and to show up contentment for the shallow and insipid thing that it is assumed to be. But ordinary human beings continue to be vulgarly high spirited. The character we all love best in Boswell is Johnson's old college companion, Mr Oliver Edwards – the man who said that he had tried in his time to be a philosopher, but had failed because cheerfulness was always breaking in.

3

I should now like to describe the new style of thinking that led to a great revival of spirits in the seventeenth century. It is closely associated with the birth of science, of course – of Science with a capital S – and the 'new philosophy' that had been spoken of since the beginning of the century referred to the beginnings of physical science; but (as I said a little earlier) we should be taking too narrow a view of things if we supposed that the instauration of science made up the whole or even the greater part of it. The new spirit is to be thought of not as scientific, but as something conducive to science; as a movement within which scientific enquiry played a necessary and proper part.

What then were the philosophic elements of the new revival (using 'philosophy' again in its homely sense)?

The seventeenth century was an age of Utopias, though Thomas More's own Utopia was already years old. The Utopias or anti-Utopias we devise today are usually set in the future, partly because the world's surface is either tenanted or known to be empty, partly because we need and assume we have time for the fulfilment of our designs. The old Utopias – Utopia itself, the New Atlantis, Christianopolis, and the City of the Sun[1] – were contemporary societies. Navi-

[1] J. V. Andreae's *Description of the Republic of Christianopolis* was first published in 1619 – see F. Held, *Christianopolis, an Ideal State of the Seventeenth Century* (Urbana, 1914); Tommaso Campanella published *The City of the Sun* in 1623 – English translation by T. W. Halliday in *Ideal Commonwealths* (London, 1885). There is an extensive literature on Utopian and chiliastic speculation, some of it rather feeble. The following are specially relevant to the idea of progress and of human improvement: J. B. Bury, *The Idea of Progress* (London, 1932); E. L. Tuveson, *Millenium and Utopia* (Berkeley, 1949); Norman Cohn, *The Pursuit of the Millenium* (New York, 1957).

gators and explorers came upon them accidentally in far-off seas. What is the meaning of the difference? One reason, of course, is that the world then still had room for undiscovered principalities, and geographical exploration itself had the symbolic significance we now associate with the great adventures of modern science. Indeed, now that outer space is coming to be our playground, we may again dream of finding ready-made Utopias out there. But this is not the most important reason. The old Utopias were not set in the future because very few people believed that there would *be* a future – an earthly future, I mean; nor was it by any means assumed that the playing-out of earthly time would improve us or increase our capabilities. On the contrary, time was running out, in fulfilment of the great Judaic tradition, and we ourselves were running down.

These thoughts suffuse the philosophic speculation of the seventeenth century until quite near its end. 'I was borne in the last Age of the World,' said John Donne,[1] and Thomas Browne speaks of himself as one whose generation was 'ordained in this setting of time'.[2] The most convincing evidence of the seriousness of this belief is to be found not in familiar literary tags, but in the dull and voluminous writings of those who, like George Hakewill,[3] repudiated the idea of human deterioration and the legend of a golden age, but had no doubt at all about the imminence of the world's end. The apocalyptic forecast was, of course, a source of strength and consolation to those who had no high ambitions for life on earth. The precise form the end of history would take had long been controversial – the New Jerusalem might be founded upon the earth itself or be inaugurated in the souls of men in heaven – but that history would come to an end had hardly been in question. Towards the end of the sixteenth century there had been some uneasy discussion of the idea that the material world might be eternal, but the thought had been a disturbing one, and had been satisfactorily explained away.[4]

[1] In a sermon delivered in Whitehall, 24 February 1625.

[2] In *Hydriotaphia*, his discourse on urn-burial. For a history of the idea of time, see Toulmin and Goodfield, op. cit. (p. 322 above, note 1).

[3] *An Apologie of the Power and Providence of God* (Oxford, 1627), an answer to Godfrey Goodman's *The Fall of Man* (London, 1616).

[4] See D. C. Allen, 'The Degeneration of Man and Renaissance Pessimism', *Studies in Philology*, **35**, 202–27, 1938.

During the seventeenth century this attitude changes. The idea of an end of history is incompatible with a new feeling about the great things human beings might achieve through their own ingenuity and exertions. The idea therefore drops quietly out of the common consciousness. It is not refuted, but merely fades away. It is true that the idea of human deterioration was expressly refuted – in England by George Hakewill but before him by Jean Bodin (by whom Hakewill was greatly influenced) and by Louis le Roy.[1] The refutation of the idea of decay did not carry with it an acceptance of the idea of progress, or anyhow of linear progress: it was a question of recognising that civilisations or cultures had their ups and downs, and went through a life cycle of degeneration and regeneration – a 'circular kind of progress', Hakewill said.

There were, however, two elements of seventeenth-century thought that imply the idea of progress even if it is not explicitly affirmed. The first was the recognition that the tempo of invention and innovation was speeding up, that the flux of history was becoming denser. In *The City of the Sun* Campanella tells us that 'his age has in it more history within a hundred years than all the world had in four thousand years before it'. He is echoing Peter Ramus:[2] 'We have seen in the space of one age a more plentiful crop of learned men and works than our predecessors saw in the previous fourteen.' By the latter half of the seventeenth century the new concept had sunk in.

The second element in the concept of futurity – in the idea that men might look forward, not only backwards or upwards – is to be found in the breathtaking thought that there was no apparent limit to human inventiveness and ingenuity. It was the notion of a perpetual *plus ultra*, that what was already known was only a tiny fraction of what remained to be discovered, so that there would always be more beyond. Bacon published his *Novum organum* at the beginning of the remarkable decade between 1620 and 1630, and had singled it out as the greatest obstacle to the growth of understanding, that 'men despair and think things impossible'. 'The human understanding is unquiet',

---

[1] Louis le Roy's remarkable work, addressed to 'all men who thinke that the future belongeth unto them', became known in England through Robert Ashley's translation of 1594 (*Of the Interchangeable Course or Variety of Things*).

[2] Cited by Hakewill, op. cit. (p. 331 above, note 3).

he wrote; 'it cannot stop or rest and still presses onwards, but in vain' – in vain, because our spirits are oppressed by 'the obscurity of nature, the shortness of life, the deceitfulness of the senses, the infirmity of judgement, the difficulty of experiment, and the like'. 'I am now therefore to speak of hope', he goes on to say, in a passage that sounds like the trumpet calls in *Fidelio*. The hope he held out was of a rebirth of learning, and with it the realisation that if men would only concentrate and direct their faculties, 'there is no difficulty that might not be overcome'. '[T]he process of Art is indefinite,' wrote Henry Power, 'and who can set a *non-ultra* to her endeavours?'[1] There is a mood of exultation and glory about this new belief in human capability and the future in which it might unfold. With Thomas Hobbes 'glorying' becomes almost a technical term: 'Joy, arising from imagination of a man's own power and ability, is that exultation of mind called glorying', he says, in *Leviathan*, and in another passage he speaks of a perseverance of delight in the continual and indefatigable generation of knowledge'.

It does not take a specially refined sensibility to see how exciting and exhilarating these new notions must have been. During the eighteenth century, of course, everybody sobers up. The idea of progress is taken for granted – but in some sense it gets out of hand, for not only will human inventions improve without limit, but so also (it is argued, though not very clearly) will human beings. It is interesting to compare the exhilaration of the seventeenth century with, say, William Godwin's magisterial tone of voice as the eighteenth century draws to an end. 'The extent of our progress in the cultivation of human knowledge is unlimited. Hence it follows . . . [t]hat human inventions . . . are susceptible of perpetual improvement.'

Can we arrest the progress of the inquiring mind? If we can, it must be by the most unmitigated despotism. Intellect has a perpetual tendency to proceed. It cannot be held back, but by a power that counteracts its genuine tendency, through every moment of its existence. Tyrannical and sanguinary must be the measures employed for this purpose. Miserable and disgustful must be the scene they produce.[2]

---

[1] *Experimental Philosophy* (London, 1664), Preface; cf. John Ray, *The Wisdom of God Manifested in the Works of the Creation* (London, 1691), pp. 124–5.

[2] William Godwin, *An Enquiry Concerning Political Justice*, 3rd ed. (London,

The seventeenth century had begun with the assumption that a powerful force would be needed to put the inventive faculty into motion; by the end of the eighteenth century it is assumed that only the application of an equally powerful force could possibly slow it down.

Before going on, it is worth asking if this conception is still acceptable – that the growth of knowledge and know-how has no intrinsic limit. We have now grown used to the idea that most ordinary or natural growth processes (the growth of organisms or populations of organisms or, for example, of cities) is not merely limited, but self-limited, that is, is slowed down and eventually brought to a standstill *as a consequence of the act of growth itself*. For one reason or another, but always for some reason, organisms cannot grow indefinitely, just as beyond a certain level of size or density a population defeats its own capacity for further growth. May not the body of knowledge also become unmanageably large, or reach such a degree of complexity that it is beyond the comprehension of the human brain? To both these questions I think the answer is 'No'. The proliferation of recorded knowledge and the seizing-up of communications pose technological problems for which technical solutions can and are being found. As to the idea that knowledge may transcend the power of the human brain: in a sense it has long done so. No one can 'understand' a radio-set or automobile in the sense of having an effective grasp of more than a fraction of the hundred technologies that enter into their manufacture. But we must not forget the additiveness of human capabilities. We work through consortia of intelligences, past as well as present. We might, of course, blow ourselves up or devise an unconditionally lethal virus, but we don't *have* to. Nothing of the kind is necessarily entailed by the growth of knowledge and understanding. I do not believe that there is any intrinsic limitation upon our ability to answer the questions that belong to the domain of natural knowledge and fall therefore within the agenda of scientific enquiry.

1797; 1st ed., 1793), vol. I, p. xxvi; vol. II, p. 535. Cf. Adam Ferguson, *An Essay on the History of Civil Society* (Edinburgh, 1767).

4

The repudiation of the concept of decay, the beginnings of a sense of the future, an affirmation of the dignity and worthiness of secular learning, the idea that human capabilities might have no upper limit, an exultant recognition of the capabilities of man – these were the seventeenth century's antidote to despondency. You may wonder why I have said nothing about the promulgation of the experimental method in science as one of the decisive intellectual movements of the day. My defence is that the origin of the experimental method has been the subject of a traditional misunderstanding, the effect of reading into the older usages of 'experiment' the very professional meaning we attach to that word today. Bacon is best described as an advocate of the *experiential* method in science – of the belief that natural knowledge was to be acquired not from authority, however venerable, nor by syllogistic exercises, however subtle, but by paying attention to the evidence of the senses, evidence from which (he believed) all deception and illusion could be stripped away. Bacon's writings form one of the roots of the English tradition of philosophic empiricism, of which the greatest spokesman was John Locke. The unique contribution of science to empirical thought lay in the idea that experience could be *stretched* in such a way as to make nature yield up information which we should otherwise have been unaware of. The word 'experiment' as it was used until the nineteenth century stood for the concept of stretched or deliberately contrived experience; for the belief that we might make nature perform according to a scenario of our own choosing instead of merely watching her own artless improvisations. An 'experiment' today is not something that merely enlarges our sensory experience. It is a critical operation of some kind that discriminates between hypotheses and therefore gives a specific direction to the flow of thought. Bacon's championship of the idea of experimentation was part of a greater intellectual movement which had a special manifestation in science without being distinctively scientific. His reputation should not, and fortunately need not, rest on his being the founder of the 'experimental method' in the modern sense.[1]

[1] See 'Induction and Intuition in Scientific Thought', pp. 73–114 above. The idea of 'stretched experience', and of the experimenter as the 'archmaster' who

Let us return to the contemporary world and discuss our misgivings about the way things are going now. No one need suppose that our present philosophic situation is unique in its character and gravity. It was partly to dispel such an illusion that I have been moving back and forth between the seventeenth century and the present day. Moods of complacency and discontent have succeeded each other during the past 400 or 500 years of European history, and our present mood of self-questioning does not represent a new and startled awareness that civilisation is coming to an end. On the contrary, the existence of these doubts is probably our best assurance that civilisation will continue.

Many of the ingredients of the seventeenth-century antidote to melancholy have lost their power to bring peace of mind today, and have become a source of anxiety in themselves. Consider the tempo of innovation. In the post-Renaissance world the feeling that inventiveness was increasing and that the whole world was on the move did much to dispel the myth of deterioration and give people confidence in human capability. Nevertheless the tempo was a pretty slow one, and technical innovation had little influence on the character of common life. A man grew up and grew old in what was still essentially the world of his childhood; it had been his father's world and it would be his children's too. Today the world changes so quickly that in growing up we take leave not just of youth but of the world we were young in. I suppose we all realise the degree to which fear and resentment of what is new is really a lament for the memories of our childhood. Dear old steam trains, we say to ourselves, but nasty diesel engines; trusty old telegraph poles but horrid pylons. Telegraph poles, as a Poet Laureate told us a good many years ago,[1] are something of a test case. Anyone who has spent part of his childhood in the countryside can remember looking up through the telegraph wires at a clouded sky and discerning the revolution of the world, or will have listened, ear to post, to the murmur of interminable conversations. For some people even the smell of telegraph poles is nostalgic, though creosote has a pretty technological smell. Telegraph poles have been assimilated into the common consciousness, and one day pylons will be, too.

'completes experience', comes from John Dee's 'Mathematical Preface' to Henry Billingsley's English translation of Euclid (London, 1570).

[1] C. Day Lewis, *A Hope for Poetry* (London, 1934), p. 107.

When the pylons are dismantled and the cables finally go underground, people will think again of those majestic catenary curves, and remind each other of how steel giants once marched across the countryside in dead silence and in single file. (What is wrong with pylons is that most of them are ugly. If only the energy spent in denouncing them had been directed towards improving their appearance, they could have been made as beautiful, even as majestic, as towers or bridges are allowed to be, and need not have looked incongruous in the country-side.)

When Bacon described himself as a trumpeter of the new philosophy, the message he proclaimed was of the virtue and dignity of scientific learning and of its power to make the world a better place to live in. I am continually surprised by the superficiality of the reasons which have led people to question those beliefs today. Many different elements enter into the movement to depreciate the services to mankind of science and technology. I have just mentioned one of them, the tempo of innovation when measured against the span of life. We wring our hands over the miscarriages of technology and take its benefactions for granted. We are dismayed by air pollution but not proportionately cheered up by, say, the virtual abolition of poliomyelitis. (Nearly 5,000 cases of poliomyelitis were recorded in England and Wales in 1957. In 1967 there were less than thirty.) There is a tendency, even a perverse willingness, to suppose that the despoliation sometimes pro-duced by technology is an inevitable and irremediable process, a trampling down of nature by the big machine. Of course it is nothing of the kind. The deterioration of the environment produced by tech-nology is a technological problem for which technology has found, is finding, and will continue to find solutions. There is, of course, a sense in which science and technology can be arraigned for devising new instruments of warfare, but another and more important sense in which it is the height of folly to blame the weapon for the crime. I would rather put it this way: in the management of our affairs we have too often been bad workmen, and like all bad workmen we blame our tools. I am all in favour of a vigorously critical attitude towards technological innovation: we should scrutinise all attempts to improve our condition and make sure that they do not in reality do us harm; but there is all the difference in the world between informed and energetic criticism and a drooping despondency that offers no remedy for the abuses it bewails.

Superimposed on all particular causes of complaint is a more general cause of dissatisfaction. Bacon's belief in the cultivation of science for the 'merit and emolument of life' has always been repugnant to those who have taken it for granted that comfort and prosperity imply spiritual impoverishment. But the real trouble nowadays has very little to do with material prosperity or technology or with our misgivings about the power of research and learning generally to make the world a better place. The real trouble is our acute sense of human failure and mismanagement, a new and specially oppressive sense of the inadequacy of men. So much was hoped of us, particularly in the eighteenth century. We were going to improve, weren't we? – and for some reason which was never made clear to us we were going to grow in moral stature as well as in general capability. Our school reports were going to get better term by term. Unfortunately they haven't done so. Every folly, every enormity that we look back on with repugnance can find its equivalent in contemporary life. Once again our intellectuals have failed us; there is a general air of misanthropy and self-contempt, of protest, but not of affirmation. There is a peculiar selfishness about modern philosophic speculation (using 'philosophy' here again in its homely or domestic sense). The philosophic universe has contracted into a neighbourhood, a suburbia of personal relationships. It is as if the classical formula of self-interest, 'I'm all right, Jack', was seeking a new context in our private, inner world.

We can obviously do better than this, and there is just one consideration that might help to take the sting out of our self-reproaches. In the melancholy reflections of the post-Renaissance era it was taken for granted that the poor old world was superannuated, that history had all but run its course and was soon coming to an end. The brave spirits who inaugurated the new science dared to believe that it was *not* too late to be ambitious, but now we must try to understand that it is a bit too early to expect our grander ambitions to be fulfilled. Today we are conscious that human history is only just beginning. There has always been room for improvement; now we know that there is time for improvement, too. For all their intelligence and dexterity – qualities we have always attached great importance to – the

higher primates (monkeys, apes and men) have not been very successful. Human beings have a history of more than 500,000 years. Only
during the past 5,000 years or thereabouts have human beings won a
reward for their special capabilities; only during the past 500 years or so
have they begun to be, in the biological sense, a success. If we imagine
the evolution of living organisms compressed into one year of cosmic
time, then the evolution of man has occupied a day. Only during the
past ten to fifteen minutes of the human day has our life on earth been
anything but precarious. Until then we might have gone under altogether or, more likely, have survived as a biological curiosity; as a
patchwork of local communities only just holding their own in a
bewildering and hostile world. Only during this past fifteen minutes
(for reasons I shall not go into, though I think they can be technically
explained) has there been progress, though, of course, it doesn't amount
to very much. We cannot point to a single definitive solution of any
one of the problems that confront us – political, economic, social or
moral, that is, having to do with the conduct of life. We are still
beginners, and for that reason may hope to improve. To deride the
hope of progress is the ultimate fatuity, the last word in poverty of
spirit and meanness of mind. There is no need to be dismayed by the
fact that we cannot yet envisage a definitive solution of our problems,
a resting-place beyond which we need not try to go. Because he
likened life to a race,[1] and defined felicity as the state of mind of those
in the front of it, Thomas Hobbes has always been thought of as the
arch materialist, the first man to uphold go-getting as a creed. But
that is a travesty of Hobbes's opinion. He was a go-getter in a sense,
but it was the going, not the getting, he extolled. As Hobbes conceived
it, the race had no finishing post. The great thing about the race was to
be in it, to be a contestant in the attempt to make the world a better
place, and it was a spiritual death he had in mind when he said that to
forsake the course is to die. 'There is no such thing as perpetual tran-
quillity of mind while we live here', he told us in *Leviathan*, 'because
life itself is but motion and can never be without desire, or without
fear, no more than without sense'; 'there can be no contentment but in
proceeding'. I agree.

[1] This simile occurs more than once in Hobbes; the passage I have in mind is
from his *Human Nature* (London, 1650), chapter 9.

INDEX

INDEX

345

Hakewill, George, 331–2

Haldane, Charlotte (formerly Burghes), 268

Haldane, John Burdon Sanderson: in nature/nurture debate, 63; and intelligence, 170–1; Clark's life of, 263–9; genetic researches, 265, 290; politics, 268

Haldane, John Scott, 264, 287–8, 306

Hales, Stephen, 101, 126

Halley, Edmund, 298–9

Halsted, William Stewart, 307

Hamerton, John L., 170

Hardy, G. H., 265

Hartley, David, 128

Harvard School of Public Health, 155

Hawkins, Harriet, 11, 17

healing (psychiatric), 57, 69n

heart: and Type A behaviour, 7, 148, 151–3

heredity: exogenetic (exosomatic or cultural), 11, 172–3, 184–7, 198; and environment, 63–4; and mental disorder, 65; Pearson on, 121; and intelligence, 169–70, 172–7, 180–2; and twin studies, 177–8; Burt and, 180; Galton on, 204; and D'Arcy Thompson's causes of form, 232–3; Teilhard and, 243; and DNA, 270; see also genetics; nature/nurture

Hering, Ewald, 288

Herrnstein, Richard, 174

Herschel, John, 123

Hertwig, R., 231

Hertzian waves, 36n

Hill, Christopher, 25

Hinde, Robert A., 192, 201

His, Wilhelm, 231

historicism, 189, 299–301

Hitler, Adolf, 72

Hobbes, Thomas, 49, 108, 191, 326, 333, 339

Hogben, Lancelot T., 63, 84n, 171, 219

Holleb, Arthur I., 165–6

Hooke, Robert, 101, 126

Hopkins, Frederick Gowland, 293

Howard, Margaret, 179

Hubble, Douglas, 142

humanities, 16, 18, 39

Hume, David, 21, 298–9

Humphrey, J. H., 164

Huxley, Aldous, 43, 45

Huxley, Julian Sorrell, 210, 220, 237, 239, 246, 248, 250–2, 277

Huxley, Thomas Henry, 13, 90, 124

Hyperboreans, 110

hypothesis, hypotheses: and induction, 33, 118, 120, 125; Newton repudiates, 45, 115, 122–3; Darwin and, 80, 121n; and scientific theory, 101–5, 116; Mill and, 104, 106, 124, 127; and deduction, 109n; connotations of word, 119, 122–5; methodological functions, 120; Bacon and, 121; as start of enquiry, 125; asymmetry (proof and disproof), 126–8; Boscovich on, 126; and discovery, 129; Whewell on, 129–30; and laws of nature, 131–2; Bernard on, 134; regulation and control of, 135; inspirational origins of, 135; 'risky', 139; and experiment, 335

hypothetico-deductive system: Popper and, 19, 34, 128–9, 131; and scientific methodology, 19–20, 101–7, 110; shortcomings, 106–8; Newton and, 123; central ideas, 125–6; explained, 129–35; and Koestler's creativity theory, 256; and Watson's discovery of DNA, 276

idealism, 187

imagination: in science, 14, 30–2, 34–7, 43–5, 48–9, 74, 108, 110, 115–16; and poetry, 38; in literature, 44, 48, 60–1; and reason, 44, 48, 124; and writing style, 49, 59; in clinical diagnosis, 100; in formulating hypotheses, 121–5, 131, 134

imbecility, 58

immunology, 4, 64–5, 186, 289

induction: shortcomings, 16–17, 86–98, 106, 120; and scientific discovery and method, 33–4, 80–2, 84, 85–8, 117; Mill and, 33, 85–6, 96, 101, 116, 119–21; and reason, 45; defined,

OXFORD PAPERBACKS

GREAT SCIENTIFIC EXPERIMENTS

ROM HARRÉ

Aristotle's study of the embryology of the chick, Theodoric's work on the causes of the rainbow, Pasteur's preparation of artificial vaccines, Jacob and Wollman's discoveries about genetics – these are among the twenty case histories that Rom Harré presents in this book. The range and intensity of human endeavour to be found throughout history in great scientific experiments make compelling reading. In *New Scientist* Peter Medawar called the book 'a great success'.

OPUS

THE PHILOSOPHIES OF SCIENCE

ROM HARRÉ

'R. Harré's *The Philosophies of Science* offers a respectably cool, hard look at scientific thought and its relationship with the great historical schools of philosophy . . . both scholarly and lucid and as good an introduction to the subject as could be wished for.' *Times Literary Supplement*

SCIENCE: GOOD, BAD AND BOGUS

MARTIN GARDNER

Martin Gardner, whose 'Mathematical Games' column has amused and puzzled readers of *Scientific American* for many years, is an unrivalled master of the art of deflating the pseudo-scientific and crackpot ideas that have run riot in the last few decades. In this book he applies his considerable wit to the examination of such phenomena as biorhythms, ESP, psychokinesis, faith healing and 'psychic surgery'.

'Absolutely fascinating'   Isaac Asimov

OXFORD PAPERBACKS

A SHORT HISTORY OF
SCIENTIFIC IDEAS TO 1900

CHARLES SINGER

This book places the basic scientific ideas developed by man in
a framework of world history, from the earliest times in
Mesopotamia and Egypt until AD 1900, and treats not only the
physical and chemical but also the biological disciplines. Published over twenty years ago to glowing reviews, it has become
a standard work.

'One reason why this new history of science is assured of an illus-
trious career is that it is a work of such consummate art . . .
masterly in conception and execution.' *New Scientist*

'this book is in the very front rank' *Advancement of Science*

'Dr Singer deserves well of Western man.' *The Economist*

OPUS

A HISTORICAL INTRODUCTION TO THE PHILOSOPHY OF SCIENCE

Second Edition

JOHN LOSEE

Since the time of Plato and Aristotle, scientists and philosophers have raised questions about the proper evaluation of scientific interpretations. John Losee's book is an account of positions that have been held on issues such as the distinction between scientific enquiry and other types of interpretation; the relationship between theories and observation reports; the evaluation of competing theories; and the nature of progress in science. Professor Losee makes the philosophy of science accessible to readers who do not have extensive knowledge of formal logic or the history of the several sciences.

A complete list of Oxford Paperbacks, including books in the World's Classics, Past Masters and OPUS series, can be obtained from the General Publicity Department, Oxford University Press, Walton Street, Oxford OX2 6DP.